*f*P

ALSO BY MICHAEL J. BEHE

Darwin's Black Box:
The Biochemical Challenge to Evolution

*The Surprising Depth of
Fine-tuning of Nature
for Life on Earth*

*The Tentative Edge of
Random Evolution*

Contingency in Biology

Laws of nature
Physical constants
Ratios of fundamental constants
Amount of matter in the universe
Speed of expansion in the universe
Properties of elements such as carbon
Properties of chemicals such as water
Location of solar system in the galaxy
Location of planet in the solar system
Origin and properties of Earth/Moon
Properties of biochemicals such as DNA
Origin of life
Cells
Genetic code
Multiprotein complexes
Molecular machines
Biological kingdoms
Developmental genetic programs
Integrated protein networks
Phyla
Cell types
Classes
Orders
Families
Genera
Species
Varieties
Individuals
Random mutations
Environmental accidents

The Edge of Evolution

The Search for the Limits of Darwinism

MICHAEL J. BEHE

FREE PRESS

New York London Toronto Sydney

FREE PRESS
A Division of Simon & Schuster, Inc.
1230 Avenue of the Americas
New York, NY 10020

FREE PRESS and colophon are trademarks
of Simon & Schuster, Inc.

For information about special discounts for bulk purchases,
please contact Simon & Schuster Special Sales:
1-800-456-6798 or business@simonandschuster.com

Book design by Ellen R. Sasahara

Manufactured in the United States of America

1   3   5   7   9   10   8   6   4   2

Library of Congress Cataloging-in-Publication Data

ISBN-13: 978-0-7432-9620-5
ISBN-10:   0-7432-9620-6

To my parents, Joseph and Helen Behe

Contents

THE ELEMENTS OF DARWINISM

Life on earth developed over billions of years by utter chance, filtered through natural selection. So says Darwinism, the most influential idea of our time. If a rare random mutation in a creature's DNA in the distant past helped the lucky mutant to leave more offspring than others of its species, then as generations passed the species as a whole would have changed. Incessant repetition of this simple process over eons built the wonders of biology from the ground up, from the intricate molecular machinery of cells up to and including the human mind.

That's the claim, at least. But is it true? To answer that question, Darwin's theory has to be sifted carefully, because it isn't just a single concept—it actually is a mixture of several unrelated, entirely separate ideas. The three most important ideas to keep straight from the start are random mutation, natural selection, and common descent.

Common descent is what most people think of when they hear the word "evolution." It is the contention that different kinds of modern creatures can trace their lineage back to a common ancestor. For example, gerbils and giraffes—two mammals—are both thought to be the descendants of a single type of creature from the far past. And so are organisms from much more widely separated categories—buffalo and buzzards, pigs and petunias, yaks and yeast.

That's certainly startling, so it's understandable that some people find the idea of common descent so astonishing that they look no further. Yet in a very strong sense the explanation of common descent is also trivial. Common descent tries to account only for the *similarities* between creatures. It says merely that certain shared features were there from the beginning—the ancestor had them. But all by itself, it doesn't try to explain how either the features or the ancestor got there in the first place, or why descendants differ. For example, rabbits and bears both have hair, so the idea of common descent says only that their ancestor had hair, too. Plants and animals both have complex cells with nuclei, so they must have inherited that feature from a common ancestor. But the questions of how or why are left hanging.

In contrast, Darwin's hypothesized mechanism of evolution— the compound concept of random mutation paired with natural selection—is decidedly more ambitious. The pairing of random mutation and natural selection tries to account for the *differences* between creatures. It tries to answer the pivotal question, What could cause such staggering transformations? How could one kind of ancestral animal develop over time into creatures as different as, say, bats and whales?

Let's tease apart that compound concept. First, consider natural selection. Like common descent, natural selection is an interesting but actually quite modest notion. By itself, the idea of natural selection says just that the more fit organisms of a species will produce more surviving offspring than the less fit. So, if the total numbers of a species stayed the same, over time the progeny of the more fit would replace the progeny of the less fit. It's hardly surprising that creatures that are somehow more fit (stronger, faster, hardier) would on average do better in nature than ones that were less fit (weaker, slower, more fragile).

By far the most critical aspect of Darwin's multifaceted theory is the role of random mutation. Almost all of what is novel and important in Darwinian thought is concentrated in this third concept. In Darwinian thinking, the only way a plant or animal becomes fitter than its relatives is by sustaining a serendipitous mutation. If the mutation makes the organism stronger, faster, or in some way hardier, then natural selection can take over from there and help make sure its offspring grow numerous. Yet until the random mutation appears, natural selection can only twiddle its thumbs.

Random mutation, natural selection, common descent—three separate ideas welded into one theory. Because of the welding of concepts, the question, Is Darwinism true? has several possible answers. One possibility, of course, is that those separate ideas—common descent, natural selection, and random mutation—could all be completely correct, and sufficient to explain evolution. Or, they could all be correct in the sense that random mutation and natural selection happen, but they might be inconsequential, unable to account for most of evolution. It's also possible that one could be wholly right while the others were totally wrong. Or one idea could be right to a greater degree while another is correct to a much lesser degree. Because they are separate ideas, evidence for each facet of Darwin's theory has to be evaluated independently. Previous generations of scientists readily discriminated among them. Many leading biologists of the late nineteenth and early twentieth centuries thought common descent was right, but that random mutation/natural selection was wrong.

In the past hundred years science has advanced enormously; what do the results of modern science show? In brief, the evidence for common descent seems compelling. The results of modern DNA sequencing experiments, undreamed of by nineteenth-century scientists like Charles Darwin, show that some distantly related organisms share apparently arbitrary features of their genes that seem to have no explanation other than that they were inherited from a distant common ancestor. Second, there's also great evidence that random mutation paired with natural selection can modify life in important ways. Third, however, there is strong evidence that random mutation is extremely limited. Now that we know the sequences of many genomes, now that we know how mutations occur, and how often, we can explore the possibilities and limits of random mutation with some degree of precision—for the first time since Darwin proposed his theory.

As we'll see throughout this book, genetic accidents can cause a degree of evolutionary change, but only a degree. As earlier generations of scientists agreed, except at life's periphery, the evidence for a pivotal role for random mutations is terrible. For a bevy of reasons having little to do with science, this crucial aspect of Darwin's theory—the power of natural selection coupled to random mutation—has been grossly oversold to the modern public.

In recent years Darwin's intellectual descendants have been aggressively pushing their idea on the public as a sort of biological theory-of-everything. Applying Darwinian principles to medicine, they claim, tells us why we get sick. Darwinian psychology explains why some men rape and some women kill their newborns. The penchant for viewing the world through Darwinian glasses has spilled over into the humanities, law, and politics. Because of the rhetorical fog that surrounds discussions of evolution, it's hard for the public to decide what is solid and what is illusory. Yet if Darwinism's grand claims are just bluster, then society is being badly misled about subjects—ranging from the cause of illnesses to the culpability of criminals—that can have serious real-world consequences.

As a theory-of-everything, Darwinism is usually presented as a take-it-or-leave-it proposition. Either accept the whole theory or decide that evolution is all hype and throw out the baby with the bath water. Both are mistakes. In dealing with an often-menacing nature, we can't afford the luxury of elevating anybody's dogmas over data. The purpose of this book is to cut through the fog, to offer a sober appraisal of what Darwinian processes can and cannot do, to find what I call *the edge of evolution*.

THE IMPORTANCE OF THE PATHWAY

On the surface, Darwin's theory of evolution is seductively simple and, unlike many theories in physics or chemistry, can be summarized succinctly with no math: In every species, there are variations. For example, one animal might be bigger than its brothers and sisters, another might be faster, another might be brighter in color. Unfortunately, not all animals that are born will survive to reproduce, because there's not enough food to go around, and there are also predators of many species. So an organism whose chance variation gives it an advantage in the struggle to survive will tend to live, prosper, and leave offspring. If Mom or Dad's useful variation is inherited by the kids, then they, too, will have a better chance of leaving more offspring. Over time, the descendants of the creature with that original, lucky mutation will dominate the population, so the species as a whole will have changed from what it was. If the scenario is repeated over and over again, then the species might eventually change into something altogether different.

At first blush, that seems pretty straightforward. Variation, selection, inheritance (in other words, random mutation, natural selection, and common descent) seem to be all it takes. In fact, when an evolutionary story is couched as abstractly as in the previous paragraph, Darwinian evolution appears almost logically necessary. As Darwinian commentators have often claimed, it just *has* to be true. If there is variation in a group of organisms, and if the variation favorably affects the odds of survival, and if the trait is inherited, then the next generation is almost certain to have more members with the favorable trait. And the next generation after that will have even more, and the next more, and the next more, until all members of the species have it. Wherever those conditions are fulfilled, wherever there is variation, selection, and inheritance, then there absolutely must be evolution.

So far, so good. But the abstract, naive logic ignores a huge piece of the puzzle. In the real world, random mutation, natural selection, and common descent might all be completely true, and yet Darwinian processes still may not be an adequate explanation of life. In order to forge the many complex structures of life, a Darwinian process would have to take numerous coherent steps, a series of beneficial mutations that successively build on each other, leading to a complex outcome. In order to do so in the real world, rather than just in our imaginations, there must be a biological route to the structure that stands a reasonable chance of success in nature. In other words, variation, selection, and inheritance will only work if there is also a smooth evolutionary *pathway* leading from biological point A to biological point B.

The question of the pathway is as critical in evolution as it is in everyday life. In everyday life, if you had to walk blindfolded from point A to point B, it would matter very much where A and B were, and what lay between. Suppose you had to walk blindfolded (and, to make the example closer to the spirit of Darwinism, blind drunk) from A to B to get some reward—say, a pot of gold. What's more, suppose in your sightless dizziness the only thought you could hold in your head was to climb higher whenever you got the chance (this mimics natural selection constantly driving a species to higher levels of fitness). On the one hand, if you just had to go from the bottom of a single enclosed stairwell to the top to reach the pot of gold, there might be little problem. On the other hand, if you had to walk blindfolded from one side of an unfamiliar city to the top of a sky-scraper on the other side—across busy streets, bypassing hazards, through doorways—you would have enormous trouble. You'd likely stagger incoherently, climb to the top of porch steps, mount car roofs, and so on, getting stuck on any one of thousands of local high points, unable to step farther up, unwilling to back down. And if, just trying to climb higher whenever possible, you had to walk blindfolded and disoriented from the plains by Lubbock, Texas, to the top of the Sears Tower in Chicago—blundering randomly over flatlands, through woods, around canyons, across rivers—neither you nor any of billions of other blindfolded, disoriented people who might try such a thing could reasonably be expected to succeed.

In everyday life, the greater the distance between points A and B, and the more rugged the intervening landscape, the bleaker are the odds for success of a blindfolded walk, even—or perhaps espe-cially—when following a simple-minded rule like "always climb higher; never back down." The same with evolution. In Darwin's day scientists were ignorant of many of the details of life, so they could reasonably hope that evolutionary pathways would turn out to be short and smooth. But now we know better. The great progress of modern science has shown that life is enormously elegant and intricate, especially at its molecular foundation. That means that Darwinian pathways to many complex features of life are quite long and rugged. The problem for Darwin, then, as with a long, blind-folded stroll outdoors, is that in a rugged evolutionary landscape, random mutation and natural selection might just keep a species staggering down genetic dead-end alleys, getting stuck on the top of small anatomical hills, or wandering aimlessly over physiological plains, never even coming close to winning the biological pot of gold at a distant biological summit. If that is the case, then random mutation/natural selection would essentially be ineffective. In fact, the striving to climb any local evolutionary hill would actively prevent all drunkards from finding the peak of a distant biological mountain.

This point is crucial: *If there is not a smooth, gradually rising, easily found evolutionary pathway leading to a biological system within a reasonable time, Darwinian processes won't work.* In this book we'll examine just how demanding a requirement that is.

A BRIEF LOOK BACK

As a practical matter, how far apart do biological points A and B have to be, and how rugged the pathway between them, before random mutation and natural selection start to become ineffective? How can we tell when that point is reached? Where in biology is a reasonable place to draw the line marking the edge of evolution?

This book answers those questions. It builds on an inquiry I began more than a decade ago with *Darwin's Black Box.* Then I argued that irreducibly complex structures—such as some stupendously intricate cellular machines—could not have evolved by random mutation and natural selection. To continue the above analogy, it was an argument that the blindfolded drunkard could not get from point A to point B, because he couldn't take just one small step at a time—he'd have to leap over canyons and rivers. The book concluded that there were at least some structures at the foundation of life that were beyond random mutation.

That conclusion stirred a lot of discussion. In particular, a lot of heat was generated in the scientific community by my inference that the structures are intelligently designed. Many people are viscerally opposed to that conclusion, for a variety of reasons. In this book, although my conclusions are ultimately the same, and will undoubtedly be opposed by some, I spend the bulk of the chapters drawing on molecular evidence, genomic research, and—above all —crucial long-term studies of evolutionary changes in single-celled organisms to test Darwinism without regard to conclusions of design. Readers who cannot accept my final conclusions should still be able to consider the evidence presented in the bulk of these chapters, before taking issue with my conclusions in the final three chapters of the book. As I will argue, mathematical probabilities and biochemical structures cannot support Darwinism's randomness, except at the margins of evolution. Still, as we seek to find the line marking the edge of randomness, there is no need to infer design.

BREAKING THE LOGJAM

Darwin's Black Box was concerned to show just that *some* elegant structures in life are beyond random mutation and natural selection. This book is much more ambitious. Here the focus is on drawing up reasonable, general guidelines to mark the edge of evolution—to decide with some precision beyond what point Darwinian explanations are unlikely to be adequate, not just for some particular structures but for general features of life. This can be compared to the job of an archeologist who discovers an ancient city buried under sand. The task of deciding whether random processes produced things like intricate paintings on walls of the city buildings (perhaps by blowing sand) is pretty easy. After all, elegant paintings aren't very likely to be made by chance processes, especially if the paintings portray not just simple geometric patterns, but images of people or animals.

But once the cherry-picking is over, the going gets tougher. Are the dark markings at the side actually a part of a painting, or just smudges? Is a pile of stones next to an exterior wall a table or an altar of some sort, or just a random collection of rocks? Is ground near the wall the remnant of a tilled field? Where lies the border of the city? Where does civilization stop and raw nature begin? Deciding on marginal cases like those is harder work, and the conclusions will necessarily be more tentative. But at the end of the study the archeologist will be left with a much clearer picture of where the city leaves off and random natural processes take over.

In a way, archeologists have it easy. Although they have to worry about the effects of physical processes on artifacts they study, they don't usually concern themselves much with biological ones. In puzzling out where might lie the far boundaries of Darwinism, uniquely biological processes of course come strongly into play. Random mutations of DNA might be likened to random accidents that befall inanimate objects. But plants and animals reproduce, stones don't. Natural selection works on living objects, not on non-living ones. Darwin's theory claims that random genetic accidents and natural selection working over eons will yield results that don't look at all like the effects of chance.

Life has been on earth for billions of years. During that time huge numbers of organisms have lived and died. Fierce struggles between different lineages over the ages are supposed by Darwinists to have led to biological "arms races"—tit-for-tat improvements of the capacity to wage biological warfare, analogous to the sophisticated twentieth-century arms race between humans in the United States and the Soviet Union. Maybe the results of those biological arms races were sophisticated living machinery, far beyond what we would ordinarily think of as the result of chance.

That's the theory. But it has proven extremely difficult to test adequately. Modern laboratory studies of random mutation/natural selection have suffered from an inability to examine really large numbers of creatures. Typically, even with heroic efforts by the best investigators, only a relative handful of organisms can be studied, only for a comparatively short amount of time, and changes in a few chosen traits are followed. At the end of such studies, while some interesting results may be at hand, it's usually impossible to generalize from them. Although scientists would love to undertake larger, more comprehensive studies, the scale of the problem is just too big. There aren't nearly enough resources available to a laboratory to perform them.

So, in lieu of definitive laboratory tests, by default most biologists work within a Darwinian framework and simply assume what cannot be demonstrated. Unfortunately, that can lead to the understandable but nonetheless corrosive intellectual habit of forgetting the difference between what is assumed and what demonstrated. Differences between widely varying kinds of organisms are automatically chalked up to random mutation and natural selection by even the most perceptive scientists, and even the most elegant of biological features is reflexively credited to Darwin's theory.

Breaking the theoretical logjam would require accurate evolutionary data at the genetic level on an enormous number of organisms that are under ceaseless pressure from natural selection. That
data simply hasn't been available in the past. Now it is.

LEAPS AND BOUNDS

Even just ten years ago any attempt to locate the edge of evolution
with any precision would have been well-nigh impossible. Too little
was known. But with the relentless march of science, especially in
the past decade, the task has become feasible.

A major difficulty of evaluating an evolutionary theory like
Darwin's has been that, while we can easily observe large changes in
animals and plants, the reasons for those changes are obscure. Darwin and other early scientists could examine, say, alterations of finch
beaks, but they couldn't tell what was causing the modifications.
Closer to our own day, mid-twentieth-century scientists could determine that some bacteria evolved resistance to antibiotics, but they
didn't know exactly what gave them that power. Only in the past half
century has science shown that visible changes are caused by mutations in invisible molecules, in DNA and proteins. The *only way* to
get a realistic understanding of what random mutation and natural
selection can actually do is to follow changes at the molecular level.
It is critical to appreciate this: Properly evaluating Darwin's theory
absolutely requires evaluating random mutation and natural selection *at the molecular level.* Unfortunately, even today such an under-
taking is intensely laborious. Yet there is no other way.

The good news is that, with much effort and insight, modern science has developed the tools to do so. A triumph of twentieth-
century science has been its elucidation of one requirement of
Darwin's theory—the underlying basis of variation. We now know
that variation in organisms depends on hidden changes in their
DNA. (For a summary of DNA structure, see Appendix A.) What's
more, scientists have catalogued myriad ways in which DNA can
change. Not only can single units (called nucleotides) of DNA accidentally change when the DNA is copied in a new generation, but

whole chunks of the double helix can accidentally either be dupli-cated or be left out. Very rarely all of the DNA in a cell is copied twice, yielding offspring with double the DNA of its parents. Other times active DNA elements resembling viruses can insert copies of themselves at new positions in the genome, sometimes dragging other bits of DNA with them. Opportunities for nature to alter an organism's DNA are virtually boundless.

Not only has the hard work of many scientists shown the under-lying basis of variation, the rate of mutation has been worked out fairly well, too. As a rule, the copying of DNA is extremely faithful. On average, a mistake is made only once for every hundred million or so nucleotides of DNA copied in a generation. But there are exceptions. In some viruses such as HIV the mutation rate is speeded up enormously.

Another critical advance in our ability to properly test Darwinism has come from DNA sequencing. In the past few decades the amount of DNA sequenced has been growing exponentially, and the number of organisms studied by sequencing has been expanded. In the mid-1990s the first complete sequence of an organism's genome—a tiny bacterium named *Hemophilus influenzae*—was published. Now the sequences of hundreds of genomes are known. Not only whole genome sequencing, but the easy ability to sequence at least key pieces of an organism's DNA gives scientists the ability to nail down the molecular changes that underlie genetic diseases, or that cause resistance to antibiotics.

Yet all that scientific progress would still not be enough to draw reasonably firm conclusions about the abilities of Darwinian evolu-tion if sufficient numbers of organisms couldn't be studied. The more organisms there are, the more opportunities random mutation has to stumble across a beneficial change and pass it on to natural selection, the firmer our conclusions about what Darwinism can do become. Studies of animals like finches can at best follow hundreds at a time. In the laboratory thousands of fruit flies might be exam-ined. That's better, but still far from enough. With thousands or even millions of organisms, a mutation comes along relatively rarely, and few of the mutations that do come along are helpful.

The natural world of course teems with organisms. There can be billions of a mammalian species on the planet at a time, such as humans or rats. In the seas there are huge numbers of fish. And these represent just the larger forms of life. There are also untold numbers of microscopic entities such as bacteria and viruses. While laboratories can't grow enough creatures to get a reasonable handle on the abilities of Darwinian evolution, nature has no such problems.

Evolution from a common ancestor, via changes in DNA, is *very* well supported. It may or may not be random. Thanks to evolution, scientists who sequence human DNA and find mutations that are helpful—against, say, our natural enemies—are not just studying the DNA of one person. They are actually observing the results of a struggle that's gone on for millennia and involved millions and millions of people. An ancestor of the modern human first sustained the helpful mutation, and her descendants outcompeted the descendants of many other humans. So the modern situation reflects an evolutionary history involving many people. When scientists sequence a genome, they are unfurling rich evidence of evolution—Darwinian or otherwise—unavailable by any other method of inquiry.

DARWINISM'S SMOKING GUN

Thanks to its enormous population size, rate of reproduction, and our knowledge of the genetics, the single best test case of Darwin's theory is the history of malaria. Much of this book will center on this disease. Many parasitic diseases afflict humanity, but historically the greatest bane has been malaria, and it is among the most thoroughly studied. For ten thousand years the mosquito-borne parasite has wreaked illness and death over vast expanses of the globe. Until a century ago humanity was ignorant of the cause of malarial fever, so no conscious defense was possible. The only way to lessen the intense, unyielding selective pressure from the parasite was through the power of random mutation. Hundreds of different mutations that confer a measure of resistance to malaria cropped up in the human genome and spread through our population by natural selection. These mutations have been touted by Darwinists as among the best, clearest examples of the abilities of Darwinian evolution.

And so they are. But, as we'll see, now that the molecular changes underlying malaria resistance have been laid bare, they tell a much different tale than Darwinists expected—a tale that highlights the incoherent flailing involved in a blind search. Malaria offers some of the best examples of Darwinian evolution, but that evidence points both to what it can, and more important what it cannot, do. Similarly, changes in the *human* genome, in response to malaria, also point to the radical limits on the efficacy of random mutation.

Because it has been studied so extensively, and because of the astronomical number of organisms involved, the evolutionary struggle between humans and our ancient nemesis malaria is the best, most reliable basis we have for forming judgments about the power of random mutation and natural selection. Few other sources of information even come close. And as we'll see, the few that do tell similar tales.

(Caveat lector: Unfortunately, in order to fully understand and appreciate the difficulties facing random mutation, and how humanity's battle with malaria illustrates them, we have to grit our teeth and immerse ourselves in details of the battle at the molecular level. I make every effort to keep technical details to a minimum, and some of them are confined to the appendices. But there is no way around the fact that this subject requires technical details.)

Although the number of malarial cells is vast, it's much less than the number of organisms that have existed on earth. Nonetheless, as I will explain, straightforward extrapolations from malaria data allow us to set tentative, reasonable limits on what to expect from random mutation, even for *all of life on earth in the past several billion years*. Not only that, but studies of the bacterium *E. coli* and HIV, the virus that causes AIDS, offer clear confirmation of the lessons to be drawn from malaria. HIV, in particular, is something of a Rosetta stone for studying random mutation, because such viruses mutate at extraordinary rates, ten-thousand times faster than the mutation rate of cells. Viruses contain much less genetic material, but it mutates so rapidly, and there are so many copies of it, that HIV alone, in just the past fifty years, has undergone more of at least some kinds of mutations than all cells have experienced since the beginning of the world.

Most of this book will focus on the operations of cells and molecules, but in the last two chapters I go further. In recent years, as science has progressed at an amazing clip, some molecular details underlying the development of different classes of animals have come to light. I make some inferences about the limits of the use of random mutation to explain features of animal life. In the final chapter I show that the conclusions reached in this book about random processes in biology mesh well with recent results from other scientific disciplines such as physics and cosmology. Together they illuminate the role of chance in nature as a whole.

GLIMMERS OF THE EDGE

One difficulty of writing a book questioning the sufficiency of Darwin's theory is that some people mistakenly conclude you're rejecting it in toto. It is time to get beyond either or thinking. Random mutation is a completely adequate explanation for some features of life, but not for others. This book looks for the line between the random and the nonrandom that defines the edge of evolution. Consider:

- On the one hand, there's malaria. An ancient scourge of humanity, in some regions of the world malaria kills half of all children before the age of five. In the middle part of the twentieth century miracle drugs were discovered that could cure the dreaded disease and hopes swelled that it could even be totally eradicated. But within a decade the malarial parasite evolved resistance to the drugs. New drugs were developed and thrown into the fight, but with only fleeting effect. Instead of humans eradicating malaria, there are worries that malaria could eradicate humans, at least in some parts of the world, as the number of deaths from the disease increased dramatically in recent years. The take-home lesson of malaria is: Evolution is relentless, brushing aside the best efforts of modern medicine.

- On the other hand, there is sickle cell disease. Although in the United States sickle cell disease is an unmitigated disaster, in Africa it shows a silver lining. It takes two copies (one from each parent) of the mutated sickle gene to get the disease. People who have just one copy do not have the disease, but they do have resistance to malaria, and they often live when others die. The gene that carries the sickle mutation arose in a human population in Africa perhaps ten thousand years ago. The mutation itself is a single, simple genetic change—nothing at all complicated. Yet despite having a thousandfold more time to deal with the sickle mutation than with modern drugs, malaria has not found a way to counter it. While the evolutionary power of malaria stymies modern medicine, a tiny genetic change in its host organism foils malaria.

- On the one hand, there's HIV. The human toll from AIDS in modern times is comparable to that from the Black Death in the Middle Ages. Modern research has developed a number of drugs to combat AIDS, but after a brief time—months, sometimes just days—they invariably lose their effectiveness. The reason is Darwinian evolution. The genome of HIV, the virus that causes AIDS, is a minute scrap of RNA, roughly one-millionth the size of the human genome. Its tiny size and rapid replication rate, as well as the huge number of copies of the virus lurking in an infected person, all combine to make it an evolutionary powerhouse. Random changes during viral replication, combined with the selective pressure exerted by medicines, allow drug-resistant varieties of HIV to prosper in a quintessentially Darwinian process. Here, evolution trumps medicine.

- On the other hand, there's *E. coli*. A normal inhabitant of the human intestinal tract, *E. coli* has also been a favorite bacterium to study in the laboratory for over a century. Its genetics and biochemistry are better understood than that of any other organism. Over the past decade *E. coli* has been the subject of the most extensive laboratory evolution study ever conducted.

Duplicating about seven times a day, the bug has been grown continuously in flasks for over thirty thousand generations. Thirty thousand generations is equivalent to about a million human-years. And what has evolution wrought?

Mostly devolution. Although some marginal details of some systems have changed, during that thirty thousand generations, the bacterium has repeatedly thrown away chunks of its genetic patrimony, including the ability to make some of the building blocks of RNA. Apparently, throwing away sophisticated but costly molecular machinery saves the bacterium energy. Nothing of remotely similar elegance has been built. The lesson of *E. coli* is that it's easier for evolution to break things than make things.

- On the one hand, there are the notothenioid fish in the Antarctic region, which can survive temperatures that should freeze their blood solid. Studies have shown that in the past ten million years tiny, incremental changes in the fishes' DNA have given them the ability to make a strange new kind of antifreeze —an antifreeze that sticks to seed crystals of ice and stops them from growing. A triumph of natural selection.

- On the other hand, there's (again) malaria. The fierce malarial parasite—the same evolutionary dynamo that shrugs off humanity's drugs—has an Achilles' heel: It won't develop in its mosquito host unless temperatures are at the very least balmy, so it's restricted mainly to the tropics. If the parasite could develop at lower temperatures it could spread more widely. But despite tens of thousands of years and a huge population size, much larger than that of Antarctic fish, it has not done so. Why can fish evolve ways to live at subfreezing temperatures while malaria can't manage to live even at merely cool temperatures?

Somewhere in the middle of such examples lies the edge of evolution.

2

ARMS RACE OR TRENCH WARFARE?

Not for nothing has malaria been nicknamed the "million-murdering death."[1] *Every year* it kills that many people—mostly young children—and sickens hundreds of times more. Human and malarial genomes have battled one another for millennia. Over the years billions of humans and astronomical numbers of malaria parasites have been at each other's throats. In their intense, enduring evolutionary struggle, any mutation that gave one an edge over the other has been favored by natural selection and has increased in number. Thanks to techniques such as DNA sequencing, many of these molecular evolutionary changes in both humans and malaria have been brought to light. Far better than Galapagos finches, pretty peppered moths, or other, more appealing examples that capture the public imagination, malaria offers our best case studies of Darwinian evolution in action.

Like some microscopic Dracula, the diabolical malaria parasite literally feeds on our blood. A single-celled organism carried by mosquitoes, it enters the bloodstream when they bite us. Once inside, malarial cells circulate until they reach the liver, stopping there for a time in order to multiply. When back in the bloodstream, a malarial cell grabs onto the surface of a human red blood cell,

seals itself tightly to it, pulls itself inside, wraps itself within a protective coating, and then starts feeding on hemoglobin. An infected
blood cell can get stuck in our veins and stop circulating. Meanwhile, the malaria inside reproduces until about twenty copies are
made. The score of new malaria cells break out of the (now trashed)
red blood cell, re-enter the bloodstream, attach to other red blood
cells, and start the process all over. Multiplying exponentially, in the
next round four hundred cells are made. In a few days a trillion new
malaria parasites can be produced and consume a large fraction of a
victim's blood.

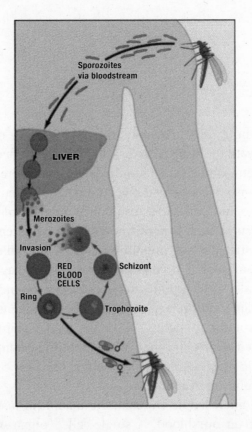

FIGURE 2.1
Invasion of red blood cells by malaria parasites. (Reprinted from Cowman, A. F.
and Crabb, B. S. 2006. Invasion of red blood cells by malaria parasites. *Cell*
124:755–66. Courtesy of Elsevier.)

Over the centuries, the human genome has tried many different defenses against malaria. In the light of modern science we now understand a great deal about each defense—not only its genetic blueprint, but often its geographical location (where on earth it has appeared) and its success at spreading through the human population. The lessons of these studies are profound and unexpected: 1) Darwinian processes are incoherent and highly constrained; and 2) the battle of predator and prey (or parasite and host), which has often been portrayed by Darwinist writers as a productive arms-race cycle of improvements on each side, is in fact a destructive cycle, more like trench warfare, where conditions deteriorate. The changes in the malaria genome are even more highly instructive, simply because of the sheer numbers of parasites involved. From them we see: 3) Like a staggering, blindfolded drunk who falls after a step or two, when more than a single tiny step is needed for an evolutionary improvement, blind random mutation is very unlikely to find it. And 4) extrapolating from the data on an enormous number of malaria parasites allows us to roughly but confidently estimate the limits of Darwinian evolution for all of life on earth over the past several billion years.

MONKEYWRENCHES

Cells are robots. Or rather, because they are so small, "nanobots." They work by unconscious, automatic mechanisms. To perform the routine tasks of their microscopic lives, cellular nanobots need sophisticated molecular machinery that works without conscious guidance. In order to stick to red blood cells, invade them, feed, and perform other essential tasks, malaria has all sorts of complicated molecular gadgets and gizmos. So does the red blood cell, for all its daily tasks.

Automated machinery of course can be quite fragile. A sophisticated mechanism can be stopped simply by some sand in its gears, or a well-placed monkeywrench. A robot navigation system can be stymied if anticipated landmarks are missing. Automated machinery may perform very well within certain limits, but can easily fail outside its working specifications.

Whenever two separate, automated mechanisms must interact, there are many opportunities for things to go wrong. Infection of a person by malaria can be pictured as the invasion of an automated city by a robot army. Although conscious humans can improvise, machines can't. If the robot army is programmed to, say, cross just one particular bridge, the invasion route can be blocked by burning that bridge. If a robot invader has a key to a certain building in the city, it can be stopped by deforming that building's lock, so the key no longer fits. On the molecular level, human resistance to malaria is much like these destructive examples.

The life of a malarial cell inside a human body is quite a complicated one. It interacts with many human structures and systems (the liver, red blood cells, walls of veins, skin, muscle, the immune system, and more) and has to perform many tasks (migration from the site of the mosquito bite to the liver, recognition of the liver, replication in the liver, attachment to a red blood cell, and so on) to successfully prepare to be sucked up in a future blood meal by a hungry mosquito. That means that the parasitic nanobot has many vulnerable points where a well-aimed monkeywrench could make the invasion grind to a halt.

The evolutionary pressure on humanity to come up with some mutational monkeywrench to counteract malaria is about as intense as it can get. If malaria were much more deadly or contagious than it is, there wouldn't be any humanity left to worry about. Any person who was born in a malarious region of the earth with some genetic change—one that, say, burned a molecular bridge—that made her resistant to the parasite would be able to have children who inherited her immunity. Her children would survive where many other children in the village would perish of the disease. When her children grew up they would be a larger fraction of the population of the village, simply because the disease kills off other children who don't have the fortunate genetic change. Over time, the descendants of the lucky woman would outnumber everyone else's descendants. Eventually every person in the village would carry the resistance mutation. Evolution by random mutation and natural selection would have changed the world, at least in that one respect.

It turns out that the above scenario has been played out hundreds of times in the course of human history. The inventive human genome has "figured out" a number of different ways to frustrate the nefarious intentions of *Plasmodium falciparum* (the most virulent species of malaria). In this chapter we'll examine several evolutionary responses to malaria by humans that show the edge of evolution is indeed past the point of many responses to parasites. Again and again, we'll see cases in which evolution is destructive, not constructive. The overriding lesson of this chapter is that the metaphor so beloved by Darwinists—that evolution is an arms race—is wrong. Evolution is trench warfare. Let's start by looking at how one well-known antimalarial monkeywrench also gums up the normal workings of the red blood cell.

HAMMERED BY SICKLE

A friendly, winsome young woman, Gail C. would stop by my laboratory about every two weeks when I was a graduate student in biochemistry at the University of Pennsylvania in the mid-1970s. Sometimes her mother would come along, too, to help Gail walk. She would step slowly and stiffly over to a table in the lab and sit down. I would draw a couple of small tubes of blood from a vein in her arm and pay her ten dollars (from my research advisor's grant money). With effort, she'd then get up and leave. I would take a bit of her blood and subject the hemoglobin to a standard laboratory procedure in which an electrical field pulls the protein through a semisolid gel. Alongside Gail's blood I'd run a drop of my own. Her hemoglobin moved somewhat more slowly through the gel than mine did—more slowly than most Americans' would—because Gail had sickle cell disease. I don't know what happened to Gail over the years, but most people with sickle cell disease die young. They suffer much pain in their shortened lives, all because their hemoglobin has a critical change in its structure.

Hemoglobin was one of the first proteins studied by scientists in modern times since it is easy to obtain (blood is full of it) and it's easy to see (most proteins are colorless, but hemoglobin is a brilliant red, and it is packed into red blood cells). Hemoglobin is the protein whose job is to carry oxygen. It isn't easy to carry oxygen—

very few proteins can do it. Hemoglobin, however, is a pro. It not only collects oxygen by binding tightly to it in the lungs, but it also dumps off the oxygen in peripheral tissues where it's needed. Although simple to describe, this little trick requires very precise engineering of the shape and amino acid sequence of hemoglobin. A number of genetic diseases are known where a single amino acid change destroys hemoglobin's ability to carry oxygen effectively.

In sickle cell hemoglobin, a single amino acid differs from normal hemoglobin. Hemoglobin has two copies of each of two distinct kinds of chains of amino acids. The four chains, two "alphas" and two "betas," all precisely stick to each other in order to do their job. In the beta chain, at position number 6 out of 146 amino acids, a single change causes trouble. That one alteration has alternately been a blessing and a curse, poison and cure, for millions of people of African descent. Although it does not significantly alter the ability of the hemoglobin to carry oxygen, it has other profound effects.

In 1904 Chicago physician James Herrick examined the blood of a young black man from Grenada, Walter Clement Noel, and was startled to see that his red blood cells were distorted.[2] Instead of the usual "Lifesaver" (or "doughnut") shape, Noel's cells displayed bizarre shapes, including crescents and sickles. The discovery of the reason for the misshapen cells took another forty years. After World War II the eminent scientist Linus Pauling first showed that the hemoglobin from people carrying the sickle cell gene moved more sluggishly than normal hemoglobin in some lab tests.[3] He correctly deduced that there was a change in the structure of the hemoglobin itself. It was the first example discovered of a molecular disease— one that is caused by an aberrant biological molecule. It took ten more years for the exact change in a single amino acid to be uncovered.[4]

Essentially, that one change causes the molecule to act as if one part were a strong magnet. Moreover, that "magnet" causes one hemoglobin to stick to a second hemoglobin, and in turn to stick to a third, and so on, until pretty much all the hemoglobin in the cell has stuck together. Sickle hemoglobin congeals into a gelatinous mess inside each red blood cell. This only happens after it has deposited its oxygen and is heading back to the heart in the veins.

FIGURE 2.2
Normal (left) and sickled (right) red blood cells. (Drawing by Celeste Behe.)

Exactly how this leads to the symptoms of the disease, including episodes of sharp pain and the death of some tissues, is still not fully understood. The most popular hypothesis has been that the stiffened cells might get stuck where the bloodstream narrows—inside tiny capillaries, whose size is often smaller than the width of a typical red blood cell. (Normal red blood cells are very flexible and easily squeeze through the capillaries.) The cells stuck at the narrows would cause a traffic jam, stopping flow through a blood vessel, which might kill cells and tissues due to lack of oxygen. That idea, however, has been disputed. We do know that the distorted shape of sickled cells is recognized as abnormal by the spleen, and the cells are destroyed more quickly than usual, leading to anemia. It's a sobering thought that, although sickle cell disease was the first molecular disorder discovered, nearly sixty years have passed and there is still virtually nothing science can do to cure it.

We also know that for people who carry the disease, having inherited the gene for it from only one parent but not both, there are nonfatal effects in the blood. About half of the hemoglobin in each of their red blood cells is the sickle form and half isn't. Usually such people have few or no health problems. It is only when two carriers mate that their children have about a one in four chance of getting the sickle cell gene from each parent, and thus inheriting the full disease. In the United States about one in ten African Americans carry the sickle trait and about a hundred thousand have the disease. In some regions of Africa close to half of the population has the trait; many have the disease.

The preceding discussion makes sickle hemoglobin sound just awful, a thoroughgoing disaster. And for many people it certainly is exactly that. But if the downside were the whole story, we'd have a real puzzle on our hands: Why has sickle cell disease persisted? Why is it so widespread? Why doesn't it disappear? Darwin's theory of evolution says that, other things being equal, those with the fittest genes will survive. But if the sickle cell gene leads to illness and death in those with two copies, why hasn't natural selection remorselessly weeded it out until none is left? The answer, of course, is that not all other things are equal. In the United States the gene is pretty much an unadulterated bane, but in Africa it can be a blessing. The sickle cell gene confers resistance to malaria.

SICK LEAVE

Thousands of years ago in malaria-ridden Africa, in a human community where many women suffered miscarriages or saw their babies die of fever, one child stayed healthy. Like the other kids, she was bitten again and again by mosquitoes, and sometimes got sick. But the illness was never severe and she quickly got back on her feet. When she grew up, she had children. Some of her children could shrug off the insect bites, too, although some couldn't, and they died. Her robust children grew up, married, and had kids of their own. As generations passed, more and more people in the region traced their ancestry back to that first healthy little girl. Let's call that first thriving child "Sickle Eve," because she became the mother of all the living who have genes for sickle hemoglobin. We should pronounce her name "Sick Leave," however, because she granted to her descendants a leave from the lethal sickness of malaria.

In one of Sickle Eve's parents, neither of whom had any special resistance to malaria, a tiny mistake happened when either the sperm or egg was made. The machinery for faithfully copying the parent's DNA, which does an almost flawless job, slipped. Instead of an exact copy, one (one!) of the billions of nucleotide components of the DNA was changed. The DNA in that reproductive cell, which would provide half of Sickle Eve's genetic information, now coded for a different amino acid (valine) at the sixth position of the second chain of hemoglobin, instead of the usual one (glutamic acid). The other half of Sickle Eve's genetic information came from her other parent, who bequeathed to her an unchanged copy of hemoglobin.

No human at the time knew why little Sickle Eve could work and play and live while other children were dying or languishing in sickbeds. But the malarial parasite knew—or found out. When a malarial cell was duly injected into Sickle Eve by the bite of a mosquito, it blithely made its usual journey to her liver and routinely changed its form. The nanobot had all its standard machinery on hand to leave the liver, recognize and stick to Sickle Eve's red blood cells, invade, feed, and reproduce. The predator docked to a red blood cell, oriented itself, released a fusillade of enzymes and proteins to prepare a tight junction, reformed its skeleton, and glided into the blood cell.

But then, from the predator's vantage point, something went terribly wrong. As the parasite fed, the inside of Sickle Eve's red blood cell changed. Random molecular motions always cause individual hemoglobins to bump into each other. But this time, instead of bouncing off as usual, they stuck together. More and more proteins clung to each other, and soon the whole liquid hemoglobin solution of the red cell began to gel. The spreading, gelatinous, semisolid mass pressed against the invader and against the red blood cell membrane, distorting its shape. As it was swept along in the bloodstream, before the parasite had time to anchor to the walls of a vein, the infected cell passed through the spleen. Ever alert to rid the body of old, damaged blood cells, the spleen grabbed the warped cell and destroyed it, along with the killer hidden inside. Sickle Eve survived, utterly oblivious to the battle her hemoglobin had fought.

TROUBLES AND TINKERING

The invisible mutation in hemoglobin, which first emerged in Sickle Eve, bestowed health upon many of her children and grandchildren. But as generations passed and her posterity grew more numerous, some descendants married other descendants. One particular husband and wife *both* had a copy of the sickle gene they had inherited from their ancestor. Their children suffered a variety of fates. Two of the couple's eight children were sickly from birth, with distended bones and spleen; they died before the age of ten. Instead of just one copy of the sickle gene, by a roll of the genetic dice these two wretched children inherited two copies. As we now know, when only half of a person's hemoglobin takes the sickle form, it won't solidify on its own. It needs a further push to make it gel. That push is supplied by the invasion of the malarial parasite. The parasite's metabolic activity raises the amount of acid in the blood cell, triggering the aggregation of the hemoglobin. For those lucky Sickle Eve descendants, only the infected cells are destroyed. But when *all* of the hemoglobin in a red blood cell is sickle hemoglobin, it needs no extra push. These children have full-fledged sickle cell disease, as opposed to the half-gene version known as "sickle trait."

Sickle cell disease is a genetic death sentence, especially in areas without access to modern medicine. But malaria is often a death sentence, too. Continuing our story from above, although two of the couple's eight children died of sickle cell disease, two others also left no descendants. One of them died of malaria and the other, crippled by the disease, never married. Those two had inherited no copies of the sickle gene, and they missed out on Sickle Eve's advantage. The four surviving children who left progeny had "sickle trait"—one copy each of normal and sickle hemoglobin genes. Over time, as the robust children married and begot their own offspring, and as other carriers of the sickle gene did likewise, "sickle trait" people flourished. This is a Darwinian success story, but it's the success of a trench-war standoff. Natural selection balanced heartbreak against heartbreak, as an equilibrium was negotiated between the plague of malaria and the curse of sickle cell disease.

How often does random mutation produce a "beneficial" change like sickle trait? By studying the DNA of many human populations, scientists have concluded that this particular mutation has arisen independently no more than a few times in the past ten thousand years—possibly only once.[5]

In evolution, equilibria are made to be broken. If, by tinkering with the machinery of life, a further mutation were able to alleviate the waste of lives from sickle cell disease without decreasing protection against malaria, then natural selection could grab hold of the variation and run with it. Over the generations that process has happened in populations of African descent, numerous times. The results can be broken down into two categories. I'll discuss the less numerous but more elegant category second, and start with the more frequent but less adroit one, something called "hereditary persistence of fetal hemoglobin," or HPFH.

As I briefly noted earlier, hemoglobin is actually made of four amino acid chains stuck together. There are two copies of one kind of chain (the alpha chain) and two copies of a similar but distinct chain (the beta chain). At least, that's the way it is in people after birth. Before birth, however, there is another kind of hemoglobin. Postnatal hemoglobin allows us to use our lungs for oxygen. But an unborn baby has to get her oxygen from her mother, through the umbilical cord. Fetal hemoglobin has a slightly different shape that allows it to pull oxygen away from Mom's hemoglobin, sort of like using a stronger magnet to pull a paperclip away from someone else's magnet. Fetal hemoglobin has two alpha chains and two gamma chains (no beta chains). While they are pretty similar to beta chains, gamma chains also have a number of differences, making the protein a stronger oxygen magnet.

Shortly before birth our bodies automatically switch from making fetal hemoglobin to making "adult" hemoglobin. But some people continue to make a noticeable amount of fetal hemoglobin throughout their lives. Their children often do the same, which is the "hereditary persistence" part of HPFH. HPFH helps sickle disease sufferers, apparently by diluting the sickle hemoglobin in their red blood cells. So instead of 100 percent sickle, folks with HPFH might have only 90 percent sickle and 10 percent fetal hemoglobin. People who have sickle cell disease but who also have HPFH often have much milder clinical symptoms than do those without HPFH. Their anemia is much less; they live longer, they can have children, so they can pass on their genes. With HPFH the execution date is often postponed, or even canceled altogether.

What causes HPFH? The DNA of a human cell codes for tens of thousands of different kinds of proteins. However, not all proteins are needed at the same time. In fact, some proteins work at cross purposes and have to be kept separate from each other. For example, after a person eats a big meal his body will normally take some of the excess sugar and turn it into starch, to be stored until energy is needed at a later time. When that time arrives, the body will break down the starch to sugar, and burn the sugar for energy. These opposing chemical processes are all catalyzed by enzymes in the cell. If all the enzymes were around and active all the time, then after a big meal the cell would be trying both to store and to burn the extra sugar, spinning its wheels. To make sure that the right proteins are made at the right times in the right order and in the right amounts, DNA contains complex "control elements"—switches that turn genes on and off. In the case of HPFH some of these control elements are broken.[6] Again, this is trench warfare. The problem (although minor) for adults with HPFH is that their hemoglobin gives them less oxygen from the air compared to normal hemoglobin. Fetal hemoglobin is not meant for adults, but if we have to break a lock or blow up a bridge to save the city, so be it.

Evolution may be trench warfare, but the armies on both sides are survivors. If a cheaper sacrifice can save a battalion, it will be more widely used because these battalions won't be as weak. A more elegant solution to the problem of the lethality of sickle cell disease is found in something called hemoglobin C-Harlem. As its name implies, C-Harlem was first discovered in a resident of New York City.[7] C-Harlem has much in common with sickle hemoglobin— both have two regular alpha chains as well as two beta chains that have the same substitute amino acid at position number 6. But the beta chains of C-Harlem also have a second mutation. Position number 73 has changed as well. That second mutation leads to surprising behavior. Half-and-half mixtures of C-Harlem with normal hemoglobin gel about as easily as the fifty-fifty mixtures of normal and sickle hemoglobin found in people with sickle cell trait. Pure sickle hemoglobin gels more strongly but pure hemoglobin C-Harlem doesn't gel at all![8] The important practical effect is that people with one normal hemoglobin gene and one C-Harlem gene have almost all the protection against malaria that Sickle Eve had. But those with two copies of C-Harlem don't have the devastating problems that people with sickle cell disease have. So C-Harlem has the advantages but not the drawbacks of sickle.

So far, the C-Harlem gene doesn't seem to have spread much. Its antimalarial properties aren't much help in contemporary New York. Its only advantage there is that it doesn't lead to sickle cell disease, but the same is true of normal hemoglobin. In Africa the C-Harlem gene would be a boon,[9] but the C-Harlem mutation apparently hasn't turned up there yet.

WHAT'S WRONG WITH THIS PICTURE?

It is crystal clear that the spread of the sickle gene is the result of Darwinian evolution—natural selection acting on random mutation. In fact, it's so transparent that the example of the sickle gene is nearly always used to teach biology students about evolution. Even in the professional literature sickle cell disease is still called, along with other mutations related to malaria, "one of the best examples of natural selection acting on the human genome."[10] No wonder—all the basic elements are there to see: the selective pressure from malaria, the single small change from the ancestral hemoglobin gene. What's more, we see additional mutations building on and modifying the first. Hereditary persistence of fetal hemoglobin (HPFH) is already widespread in Africa, ameliorating the problems of the sickle gene. The C-Harlem gene, which builds directly on the foundation of the sickle gene and would entirely eliminate the drawbacks of the sickle mutation, has not yet turned up in Africa, where it would do the most good, but there's little doubt that over time it, or something like it, will appear. Perhaps, as advocates of Darwinian evolution argue, we can jump directly from this pristine example to the conclusion that all of life—the complex machinery of the cell, the human mind, and everything in between—can be explained the same way.

But can we? The defense of vertebrates from invasion by microscopic predators is the job of the immune system, yet hemoglobin is not part of the immune system. Hemoglobin's main job is as part of the respiratory system, to carry oxygen to tissues. Using hemoglobin to fight off malaria is an act of utter desperation, like using a TV set to plug a hole in the Hoover Dam. Even leaving aside the question of where the dam and TV set came from—which is no small question—it must be conceded that this Darwinian process is a tradeoff of least-bad alternatives. The army in its trenches is suffering loss upon loss. No matter which way it turns, in this war fought by random mutation and natural selection, it is losing function, not gaining.

Sickle hemoglobin is not the only change that malaria has wrought in the human genome. Let's explore a handful of others—changes that have literally been written in the blood of many humans—some of which have arisen independently hundreds of times over the past ten thousand years. Let's see if the picture of random mutation we get from sickle hemoglobin is an exception or the rule.

SIEVE

Since the primary target of malaria is the red blood cell, it's not surprising that hemoglobin has endured a number of evolutionary changes. Besides the sickle cell mutation, other changes to hemoglobin have also arisen that slow the parasite's progress. One change that's similar to sickle hemoglobin—but with illuminating differences—is something called hemoglobin C.[11] Confusingly, hemoglobin C isn't the same thing as hemoglobin C-Harlem. Like sickle cell hemoglobin, hemoglobin C (abbreviated HbC) has just one difference from normal hemoglobin in its amino acid sequence. Again like sickle cell hemoglobin, the change occurs in the sixth position of the beta chain. But in the case of hemoglobin C the substitution is of a close relative. Like the amino acid it replaces, the new one is electrically charged. The difference between them is that the new one is positively charged, whereas its predecessor is negatively charged.

Hemoglobin C is not as widespread as sickle hemoglobin, but it does occur frequently in some regions of west Africa. Unlike sickle hemoglobin, HbC does not solidify, so it doesn't seem to cause any major problems itself, certainly none as severe as sickle cell disease. Nonetheless, it seems to help people fight malaria. We aren't quite sure why, but experiments indicate that HbC is less sturdy than normal hemoglobin. When the malarial parasite enters the red blood cell, the increased stress inside the cell apparently causes the mutant hemoglobin to unfold more readily, exposing the parasite to reactive oxygen molecules that may damage it. The unfolded hemoglobin may also indirectly cause the cell to be destroyed by the spleen.

HbC gives protection from malaria and doesn't cause nearly as many problems as sickle does. Yet HbC hasn't spread throughout Africa, replacing sickle hemoglobin. Why not? The answer lies with the folks who get one gene for hemoglobin C from one parent, but a normal gene for regular hemoglobin from the other parent—the "heterozygotes." Although only one sickle gene gives excellent protection against malaria, one hemoglobin C gene gives only a small amount of protection. To get a full dose of protective power, a person has to have two copies of the C gene [12]; in other words, she has to inherit one from each parent. Ironically, while two copies of the sickle gene kill, a double dose of the C gene cures. On the other hand, while a single sickle gene cures, a single C gene doesn't do much.

To see why hemoglobin C is limited to a few regions of Africa, let's contrast the fates of Sickle Eve, the first person to have the mutant sickle gene, and another little girl I'll call, inelegantly, "C-Eve," who was the first to have the mutant C gene. If Sickle Eve is pronounced "Sick Leave," let's stretch a bit and pronounce C-Eve as "Sieve," because all too often protection from malaria trickles through her grasp.

When Sickle Eve was born she flourished, shrugging off the mosquito bites that sickened and killed other children in her village. About half of her own children inherited her immunity, and their progeny quickly became more numerous than that of other villagers. Only later, when the sickle gene became common enough for a husband and wife to each have a copy, did problems arise, as some children inherited two sickle genes and thus sickle cell disease. In contrast, when C-Eve was born in a neighboring village she was no better off than most other kids in the region. As a toddler she was constantly being bitten by malarious mosquitoes. Like many other kids, she developed fever and was often desperately sick. But, as luck would have it, she was part of the fraction of children who survived up to age five or so, where the threat of death from malaria greatly diminishes.[13] C-Eve grew up, married, and watched in helpless agony as half of her children died, either through miscarriage or by fever, as infants and toddlers.

But the other half of C-Eve's children survived, grew up, and had kids of their own. When some of these descendants moved to Sickle Eve's ancestral village and took spouses there, their children were much worse off from malaria than many other children in the village—all descendants of Sickle Eve—so their line quickly died out in that village. However, the descendants who stayed in C-Eve's ancestral village became somewhat more numerous than the descendants of others in that village. One-half of C-Eve's children carried the mutation (to little effect), and when the right descendants had children, one-quarter of those kids both had mutations and were much more likely to survive. Still, when *their* kids grew up and married, often their children would not have nearly as much resistance as their mother, unlike the children of Sickle Eve. Over time, though, more and more lucky babies were born in the village, some of whom married each other, and their children always had strong resistance. But there was a catch. When C-Eve's offspring moved to a different village and married a local boy or girl, their children lost resistance. So for C-Eve's progeny to prosper, they had to stay close to home and, like Charles Darwin, marry their kissing cousins.[14] (Meanwhile, Sickle Eve's children could spread their advantage far and wide.)

With the advance of science we can now understand the reasons behind these seemingly arbitrary twists of fate, which would certainly have baffled C-Eve and her descendants. Since the sickle gene gives resistance to malaria with just one copy, Sickle Eve prospered from the beginning, as did many of her descendants until they married each other and some children inherited two copies of the sickle gene. However, since the hemoglobin C gene needs two copies to be effective, and gives only a small amount or malaria resistance in single copy, then C-Eve was no better off than her fellow villagers. As C-Eve's progeny increased—initially just by luck—as a percentage of the population of the village, then the small amount of protection from a single copy of the C gene started to give them a statistically better chance of surviving than those with no copy, so the C-gene started to take hold.

As more villagers had the C gene, there was a better chance that two of them would marry, and have at least a few children who had two copies, and thus full protection against malaria. When those healthy kids grew up and married another villager, it was still rather likely that most or all of their children would have only a single C gene. But if those fortunate children moved to another village that had no one with a C gene, then the children of the intervillage marriage would necessarily have only a single C gene, and so lose almost all the protection from malaria their parent had.[15] So to prosper, the children had to stay close to home and preferably marry close relatives.

The good part of hemoglobin C is that people with two mutant genes have few health problems, unlike either people with two sickle genes (who have sickle disease) or people with two normal hemoglobin genes (who are vulnerable to malaria). The downside is that the C gene spreads only very slowly. In a head-to-head contest the C gene should replace the sickle gene in endemic malarial regions over enough time, all other things being equal, because at C's best it does as well as the sickle gene at resisting malaria, without the severe collateral damage. However, since all things are rarely ever equal, the prediction is far from certain.[16]

TAKE-HOME LESSONS

Let's pause here for a moment to consider several simple points about the sickle and HbC mutations. The first point is that both sickle and HbC are quintessentially hurtful mutations because they diminish the functioning of the human body. Both induce anemia and other detrimental effects. In happier times they would never gain a foothold in human populations. But in desperate times, when an invasion threatens the city, it can be better in the short run to burn a bridge to keep the enemy out.

A second point is that the mutations are not in the process of joining to build a more complex, interactive biochemical system. The sickle and C mutations are mutually exclusive, vying for the same site on hemoglobin—the sixth position of the beta chain. They do not fit together to do something. A related point is that neither hemoglobin mutation occurs in the immune system, the system that is generally responsible for defending the body from microscopic predators. So the mutations are neither making a new system nor even adding to an established one. In this book we are concerned with how machinery can be built. To build a complex machine many different pieces have to be brought together and fitted to one another.

A final, important point is that even with just those two simple mutations the process is convoluted almost to the point of incoherence.[17] Even with just the sickle and C genes—with heterozygote versus homozygote advantage and with varying detrimental effects—the interplay of the mutant and normal genes is chaotic and tangled. Sickle is better in the beginning but C is better in the end; sickle spreads quickly, establishing itself as king of the hill before C can get started; sickle trait carriers are better off marrying someone outside the clan, but C carriers do better by marrying relatives; and so on. It's not hard to imagine a few more mutations popping up in hemoglobin or other genes to make the process truly Byzantine in its intricacy and cross-purposedness. The chaotic interplay of genes is not constructive at all. In the everyday world of our experience, when many unrelated threads get tangled together, the result is not a pretty tapestry—it's a Gordian knot. Is that where Darwinian evolution also leads?

MAN OVERBOARD

The mutations that yield sickle hemoglobin and HbC are both subtle. In each case a single alteration in a specific nucleotide—one of the building blocks of DNA—altered the gene for the beta chain of hemoglobin so that only one amino acid was different in the mutant proteins. There are, however, cruder, more drastic mutations that also aid in the war with malaria. In this set of mutations a whole gene is tossed out—either accidentally deleted or altered so that it no longer produces any working protein. When that protein is hemoglobin, the resultant class of conditions is called "thalassemia," from the Greek word for the sea, because it was first noticed in people who lived by the Mediterranean Sea. In fact, thalassemia is widespread in Africa, the Middle East, and Asia.

Healthy hemoglobins have four chains—two alpha chains and two beta chains. In a person with thalassemia, however, a copy of a gene for one of the kinds of chains of hemoglobin is either deleted or switched off. This causes an imbalance in the total amount of chains that are made by the cell. In some thalassemias there is an excess of beta chains; in others there is an excess of alpha chains. Thalassemias in which the alpha chain is in short supply usually lead to less severe anemia than when the amount of beta chain is deficient. In most alpha thalassemias, a whole gene is deleted. The effects of the deletion can vary considerably. Normal persons have four alpha genes, inheriting two from each parent. Alpha-thalassemic children of alpha-thalassemic parents can be missing one, two, three, or all four alpha genes. If only one or two alpha genes are missing, the remaining two or three alpha genes apparently make enough of the alpha chain to supply the red blood cell with enough working hemoglobin to get by. If no alpha genes are present, the child dies before birth.

Sickle hemoglobin and hemoglobin C are very specific mutations, each caused by one particular amino acid. In contrast, many—about a hundred—different kinds of mutations halt production of the beta chain, resulting in thalassemias.[18] Sometimes a whole beta gene is deleted. Other times the mutant gene has lost important processing signals, which leads either to a malformed beta chain or to a decrease in the amount of beta chain the gene can make. In other cases a few or even just a single nucleotide is changed in the beta gene, rendering it completely nonfunctional. Because there are so many different mutations that can cause thalassemia versus just one that can make sickle or hemoglobin C, thalassemias originate by chance much more frequently. They spread quickly in malarious areas before particular mutations like sickle have a chance to even get started.

Thalassemia is another detrimental mutation, like sickle hemoglobin or HbC. Even in its mildest form, it is a diminishment of the functioning of the system that supplies oxygen to the tissues. But thalassemia is useful in slowing a malarial invasion. Studies have shown that, although thalassemia doesn't protect nearly as well as one copy of a sickle cell gene, it still gives about 50 percent protection against malaria (at least for one type of thalassemia), probably by making the red cell more fragile.[19] It's a bridge that can be burned to thwart malarial attack.

OTHER RED CELL GENES

Hemoglobin is a good protein to alter in the fight against malaria because it's the most abundant protein in the red blood cell. If hemoglobin isn't working just right—if it gels or is unstable—then the red cell in which malaria travels will not be as strong, and will have a shortened lifespan. Any additional stress on the fragile cell from the malarial parasite might quickly push it over the edge, causing it and the parasite to be destroyed by the spleen. It's not surprising that so many different mutations to hemoglobin have prospered in malarious areas, since there are many different ways to foul up the workings of a machine.

But although hemoglobin is the most abundant protein in the red blood cell, it certainly isn't the only one. When some of the other red blood cell proteins mutate, a person acquires some resistance to malaria. In this section I'll briefly mention several of those proteins. The point to keep in mind is this: As with hemoglobin, the mutations all involve diminishing the function of a protein, or jettisoning it altogether. Readers who do not feel it necessary to pay close attention to the technical details may prefer to skip the rest of this section.

One useful gadget of the red blood cell nanobot is a protein called "glucose-6-phosphate dehydrogenase," which (mercifully) can be abbreviated as G6PD. G6PD is responsible for generating "reducing power" in the cell, which can be thought of as something akin to antacid. The red blood cell has a dangerous job. A cell carrying a lot of oxygen can be likened to a person carrying glass bottles of acid. Once in a while one of those bottles is going to accidentally drop and break, and the person will be splashed and burned. In the red blood cell the oxygen is like the acid and the hemoglobin like the glass bottles. Although hemoglobin does a good job, even in the best of circumstances, occasionally a hemoglobin molecule "breaks," the oxygen (or chemically related material) escapes, and the cell can be burned. To deal with the anticipated breakage, G6PD leads to a chemical (called glutathione) that, under the guidance of other repair machinery, sops up the spilled oxygen, limiting the damage as much as possible.

Junking G6PD makes the red blood cell more fragile, which as we have seen can be a net plus during an invasion by malaria.[20] Infected, more-fragile cells may be spotted more easily by the spleen and destroyed. Hundreds of mutations are known that alter the amino acid sequence of G6PD, and either destroy or greatly diminish its effectiveness. Depending on the nature of the mutation, and on whether it occurs in a man or woman,[21] it can lead to anemia. Because so many different mutations can break the G6PD gene, the rate of their appearance is much, much higher than the rate of appearance of the mutation for sickle hemoglobin. G6PD mutations are widespread in malarious regions around the world, from Africa to Asia. Studies have indicated they can give roughly the same degree of protection against malaria as thalassemia.[22]

Another protein machine that normally helps keep the red blood cell nanobot humming is called "band 3" protein. Band 3 protein is situated in the membrane of the red blood cell. Its job is to be a supply-exchange portal, allowing some kinds of needed materials to come into the cell, and to pump out waste products.[23] In some malarious regions, particularly in Melanesia, the population contains a high percentage of people with defective band 3 genes, which again seems to confer some resistance to malaria.[24] How it does so is not clear. Perhaps it, too, works by increasing the fragility of the red blood cell. Alternatively, it may work by forming clumps more easily than usual. Clumped band 3 proteins are normally a sign that a cell is aging, and the body targets those cells for destruction.[25] In some regions the percentage of the population with one copy of a defective band 3 gene is quite high (about 20 percent) but no people have been found there who have two defective copies. The grim implication is that inheriting two broken copies kills a child before birth. A high price for a population to pay, but apparently less than enduring the full brunt of malaria.

Malaria grabs onto red blood cells by seizing on certain proteins that are akin to antennas on the outside of the cells. One "antenna" protein sticking out from the surface of the red blood cell is called "Duffy antigen." A species of malarial parasite called *P. vivax* (a milder cousin of the vicious *P. falciparum*) specifically grabs hold of it as a prelude to invasion. However, almost all people in west and central Africa are completely immune to *P. vivax* malaria because their red blood cells no longer make Duffy antigen. A single nucleotide change in DNA does the trick.[26] Like HbC, the mutated gene has to be present in two copies to have much of a benefit against malaria. Turning off Duffy antigen in red cells has little noticeable ill effect, although it may be associated with an increased risk of prostate cancer.

Some other mutations are also known to help in the fight against malaria, but the ones discussed so far are the best characterized. Here's the bottom line: They are all damaging. Some are worse than others, but all are diminishments; none are constructive. Like sickle hemoglobin, they are all acts of desperation to stave off an invader.

TABLE 2.1
Human genetic effects of malaria trench warfare

"This burden is composed not only of the direct effects of malaria but also of the great legacy of debilitating, and sometimes lethal, inherited diseases that have been selected under its impact in the past" (Carter and Mendis, 2002, p. 589).

Gene	Mutation	Adverse effects (clinical / molecular)
Hemoglobin, beta chain	HbS: specific point mutation	anemia; usually lethal in two copies / increased fragility of red blood cell
	HbC: specific point mutation	slight anemia / increased fragility of red blood cell
	HbE: specific point mutation	none apparent
	thalassemia: various point mutations, deletions	anemia / broken gene
Hemoglobin, alpha chain	thalassemia: deletion	anemia / broken gene
Hemoglobin, gamma chain	HPFH: various deletions, point mutations in control regions	none apparent / broken genetic controls
G6PD	various point mutations, deletions	anemia / decrease or loss of G6PD function
Band 3 protein	deletion	lethal in two copies / broken gene
Duffy antigen	specific pont mutation in control region	none apparent / protein no longer expressed in red blood cells

BACK TO THE STONE AGE

Over much of the second half of the twentieth century the United States and Soviet Union engaged in mutual saber-rattling of the most unnerving sort. Verbal threats were backed up by a terrifying arms race. The United States developed nuclear weapons, then so did the U.S.S.R. One side made larger weapons; the other improved the accuracy of theirs. One side placed weapons in countries close to the other; the other side put them in submarines. One side developed an antimissile system; the other invented sophisticated evasion equipment. By some miracle the weapons haven't been used yet, but there are no guarantees for the future.

The arms race in ballistic missiles and related technology between the human superpowers was of course carried out by intelligent agents acting purposefully to achieve a goal (however misconceived). That is pretty much the absolute opposite of Darwinian evolution, which posits only blind, purposeless genetic accidents, some of which might be favored by the automatic effects of natural selection. Nonetheless, some Darwinists have professed to see in human arms races a good analogy for blind evolution. For example: Suppose that in the distant past the ancestor of a modern cheetah started hunting the ancestor of a modern gazelle. At that time both were relatively slow compared to their descendants, but even then some of the faster cheetahs caught some of the slower gazelles. After dinner, pairs of faster, better-fed cheetahs repaired to the brush to produce more offspring than hungry, slower cheetahs; laggard gazelles simply disappeared into cheetah stomachs while the speedier ones survived. Natural selection in action.

The occasional random mutation that made a cheetah or a gazelle a bit faster would favor its descendants, so, the story continues, speedier cheetahs would set the stage for the evolution of speedier gazelles, and vice versa. Over many generations both cheetahs and gazelles would get faster, even though the average number of gazelles consumed by cheetahs might stay roughly constant. So by competing against each other, the two species would both get better, although neither would entirely surpass the other. Another label that has been pasted on this concept is the "Red Queen effect,"

after the silly statement by the Red Queen to Alice that in Wonder-
land you have to run as fast as you can just to stay in the same place.
The idea is that in evolution, a species and its enemies all have to
keep getting better just to keep surviving.

At first blush the idea of an arms race sounds plausible, and some
ardent Darwinists have proclaimed it to be perhaps the most important factor in progressive evolution—the building of coherent, complex systems. In his classic book defending Darwin, *The Blind
Watchmaker*, Oxford biologist Richard Dawkins announced:

> I regard arms races as of the utmost importance because it is largely
> arms races that have injected such "progressiveness" as there is in
> evolution. For, contrary to earlier prejudices, there is nothing
> inherently progressive about evolution.[27]

And:

> The arms-race idea remains by far the most satisfactory explanation
> for the existence of the advanced and complex machinery that ani-
> mals and plants possess.[28]

Dawkins's deduction of the importance of arms races in evolution is
wishful thinking. To play along, let's consider the illustration of the
cheetah and gazelle, but a bit more skeptically. How could a gazelle
better avoid a faster cheetah? One way, as the standard story has it, is
to become faster itself. But another way might be to become better at
making quick turns, in order to dodge the predator in a chase. Or to
develop stronger horns for defense. Or tougher skin. Or grow bigger. Or develop camouflage. Or graze where cheetahs aren't. Or
when cheetahs are asleep. Or close to a forest in which to hide. Or
any of a hundred other strategies. Or all of the above. Like the many
different ways human genes can change to help fend off malaria,
gazelles could change in numerous, unconnected ways.

The Just-So story seems plausible at first only because it
doggedly focuses its gaze on just one trait—speed—ignoring the
rest of the universe of possibilities. But in the real world Darwinian
evolution has no gaze to focus; it is blind. In a blind process, there
can be no intentional building on a single trait, continually improving a discrete feature. Anything that works at the moment, for the

moment, will be selected whether it is "progressive" or not—to hell with the future. The descendants of a slightly faster gazelle might go on to develop slightly better camouflage or slightly different feeding strategies or to slightly change any of innumerable other traits, eliminating the need for speed. If that were the case, gazelles would not keep getting faster. They would change over time in myriad, disjointed, jumbled ways. There is no reason to expect the coherent development of a single trait in a Darwinian arms race.

You may be wondering whom to believe at this point, since I am just countering Dawkins's suppositions with my own. But consider this. Although there have been some studies showing modest arms races with smaller animals—ants, other invertebrates, and microorganisms—there have been absolutely no studies that document that large animals change in the way Dawkins supposes. We know the most about "arms races" between parasites and hosts. Far and away the most extensive relevant data we have on the subject of evolution's effects on competing organisms is that accumulated on interactions between humans and our parasites. As with the example of malaria, the data show trench warfare, with acts of desperate destruction, not arms races, with mutual improvements.

The thrust and parry of human-malaria evolution did not build anything—it only destroyed things. Jettisoning G6PD wrecks, it does not construct. Throwing away band 3 protein does likewise. Sickle hemoglobin itself is not an advancement of the immune system; it's a regression of the red blood cell. Even the breaking of the normal controls in HPFH doesn't build a new system; it's just plugging another hole in the dike.

The arms race metaphor itself is misconceived. The relationship between malaria and humans is nature red in tooth and claw. Real arms races are run by highly intelligent, bespectacled engineers in glass offices thoughtfully designing shiny weapons on modern computers. But there's no thinking in the mud and cold of nature's trenches. At best, weapons thrown together amidst the explosions and confusion of smoky battlefields are tiny variations on old ones, held together by chewing gum. If they don't work, then something else is thrown at the enemy, including the kitchen sink—there's nothing "progressive" about that. At its usual worst, trench warfare is fought by attrition. If the enemy can be stopped or slowed by burning your own bridges and bombing your own radio towers and oil refineries, then away they go. Darwinian trench warfare does not lead to progress—it leads back to the Stone Age.

In a real war, everything relentlessly gets worse. In its real war with malaria, the human genome has only diminished.

THE MORE THE MERRIER

Vain creatures that we are, no topic holds our interest more than ourselves. Yet perhaps a focus on us humans distorts the picture. Although our cities seem crowded, the number of humans on earth is actually minuscule compared to the numbers of microscopic creatures. For each human sick with malaria in the world, there are roughly a trillion parasites. The more individuals of a species there are, and the shorter the life span of each generation, the more opportunities for beneficial mutations pop up. To get a better idea of what random mutation and natural selection—Darwinian evolution—can do, let's consider that far more numerous species, *Plasmodium falciparum* itself. How has the million-murdering death evolved during its encounter with humans?

3

THE MATHEMATICAL LIMITS
OF DARWINISM

For millennia humans struggled with malaria in an unconscious war, where the only defense was by attrition in the evolutionary process of random mutation and natural selection envisioned by Charles Darwin. But in the past five hundred years a radically different factor has transformed the war. Using our ability to reason, over time we humans have learned much about the world that was hidden from our ancestors. In particular, the discovery of microscopic predators has allowed us to take the fight to the enemy. Rather than waiting for a lucky mutation to come along, medicines have been both discovered and invented that can kill malaria. At first the new, rational phase of the war was restricted to the use of plants that nature herself provided. But in the past three-quarters of a century advances in chemistry, medicine, and basic biology have led to new drugs that nature never thought of.

The initial glorious result was sweeping victory wherever the battle was joined. The cruel malarial parasites perished by the uncounted trillions. In the giddy days of the 1950s there was much talk of totally eradicating malaria. Humanity would soon live in a world free of its ancient nemesis. Optimism was cheap. Around the same time it was thought that other tiny scourges—virulent bacteria, viruses, and even agricultural insect pests—could be fended off by drugs and insecticides. Early victories on those fronts were also easy to come by.

But the mood these days is somber. The miracle drugs are in retreat or have failed. The title of a recent article in the journal *Science,* "A Requiem for Chloroquine,"[1] refers to the medicine that was for decades the standard treatment for malaria. Gone is talk of a final victory over malaria, replaced by modest hopes that maybe it at least can be contained. Malaria seems actually to be on the increase in Africa.[2] Although it does not yet look as if we're headed back to the bad old days, the war between humanity and *P. falciparum* has reached an uneasy stalemate. To our chagrin, the unexpected stubbornness of the parasite proves that evolution is powerful for foes and friends alike. Sickle Eve isn't the only one to benefit from a fortunate mutation. *P. falciparum* knows that trick, too. Here, too, lies some of the best available evidence for Darwinism—as well as clear evidence of its limits.

A NATURAL CURE

For centuries of recorded history, even while entire civilizations were obliterated by it, humanity remained helpless in the face of malaria. Hippocrates—the Father of Medicine himself—ascribed its periodic fevers to an imbalance in the body's four "humors" (blood, phlegm, and black and yellow bile). Medieval men of medicine thought that the fevers were caused by bad air (*mal'aria* is Italian for bad air) rising from fetid swamps in the summer. Treatments included bleeding with leeches—in retrospect not a good idea for people suffering from malaria-induced anemia.

The modern fight against malaria began with the discovery that powder made from the bark of the cinchona tree in the South American Andes was useful for treating fever. Although it probably wasn't used for malaria by the local natives, in the seventeenth century European settlers brought it back across the Atlantic, where it was first used unwittingly on malarial patients. When a few members of royalty were cured by cinchona bark, demand soared. The cinchona tree was cultivated for export by the Dutch on their Indonesian colonial plantations. The bark became widely available and literally changed the world. With quinine from cinchona bark, Europeans could colonize and operate commercial ventures in tropical climates, usually with the help of African or Indian workers who, because of sickle trait or thalassemia, had a measure of natural resistance to malaria.

Quinine, the active ingredient, was first isolated from cinchona bark by French chemists in the early nineteenth century. It was not until the 1940s, however, that the eminent organic chemist Robert Woodward synthesized the compound in his laboratory at Harvard. In the 1930s a synthetic antimalarial drug similar to quinine was developed by a German pharmaceutical firm. During World War II a cache of the drug was captured by the American army, which reformulated it as chloroquine. Like quinine, chloroquine is a rather small molecule (not a big, "macro" molecule) and has a core structure called a "quinoline." Chloroquine is a simpler molecule than quinine and thus is much easier to synthesize in the lab. Because of its effectiveness and cheap production cost, chloroquine became the drug of choice for the treatment of malaria for decades.

Yet within a few years after the introduction of chloroquine, reports popped up of its failure to cure some patients. As time passed, the reports became more frequent, and by the 1980s chloroquine was ineffective against the majority of cases of P. falciparum. To understand how chloroquine once worked and why it has failed, let's look at some of the nuts and bolts of the malarial nanobot.

TOXIC WASTE

P. falciparum feeds on the hemoglobin inside a human red blood cell. But there's a hitch. Although the parasite can digest the protein part of hemoglobin—breaking it down to amino acids which it reuses to help construct copies of itself—it can't use the heme part of hemoglobin. Heme, which gives blood its red color, is a small molecule (roughly the same size as chloroquine, and about one-thirtieth the size of one of the four protein chains of hemoglobin) that sticks to the protein, but heme is not made of amino acids. It is an indigestible and poisonous waste product that the parasite urgently has to neutralize. If heme accumulates in the parasite's digestive compartment—its "stomach"—the bug dies. Normally, *P. falciparum* ties together the waste heme to form something called hemozoin, which is harmless. But chloroquine interferes with waste removal, so that the toxic heme remains free. Exactly how heme kills the parasite isn't quite clear. But the fate of drowning in its own waste is a fitting end for the agent of such human misery.

For chloroquine to kill malaria, it has to get inside the parasite's "stomach" and stay there for a while. In fact, the parasite itself grabs the drug and concentrates it ten-thousand-fold in its digestive vacuole. The process is complex, and there are many ways it could short-circuit, yielding resistance to the drug. Let's speculate about the possibilities, without worrying yet about the mutational complexity.

Perhaps the digestive vacuole's environment could be changed a bit to make it less congenial to the drug. Or possibly a protein pump that ordinarily removes other things from the cell could be altered a bit, to toss out the poison. Or maybe some of the ordinary repair machinery of the cell could be tweaked to chemically damage the chloroquine. Or possibly *P. falciparum* could change some of the components of its membrane to stop the entry of the chemical. Or it might develop an alternative way to deal with waste. The large number of potential ways for the parasite to counter chloroquine makes it difficult for scientists to track down what really is going on in drug resistance. It has only been in the past few years that the mutation that makes *P. falciparum* resistant to chloroquine has been unmasked. Even now researchers are unsure if it's the whole story, but they are confident that at least it is a large chunk of the story.

A group led by Thomas Wellems of the National Institutes of Health made the discovery through a series of genetic studies of the parasite. First they narrowed it down to just one of the parasite's fourteen chromosomes. More sophisticated (and laborious) studies further narrowed the region containing the resistance gene to a four-hundred-thousand-nucleotide region of that chromosome,[3] and then eventually to a thirty-six-thousand-nucleotide region.[4] Painstaking analysis of the details of this area of the parasite's genetic map uncovered a previously unnoticed gene.

When that gene was sequenced, workers were able to determine the amino acids in the protein they were seeking. It was a needle in a haystack: one of the approximately fifty-three hundred proteins that the parasite's DNA encodes.[5] With the progress that biology has made over the past few decades, scientists are able to tell a great deal about what role a protein is likely to play in a cell just by looking at its amino acid sequence, even before conducting any experiments. They have learned to recognize patterns in amino acid sequences that are reliably found in proteins that do particular kinds of jobs. As an analogy, if an engineer saw an unfamiliar machine that had wheels, he would guess that it was probably used for transportation of some sort; if the machine had a sharp blade, that part probably was used for cutting something; and so on.

The sequence of the protein—dubbed PfCRT, for *P. falciparum* chloroquine resistance trait—revealed that the protein contained ten separate stretches of amino acids that were all hydrophobic (water-hating), or oily. This suggested that in the cell, those regions would be stuck in a membrane, which is itself made of oily molecules that prefer contact with other oily molecules. The other regions of PfCRT were not so hydrophobic and probably stuck out into the water on one side of the membrane or the other. Other proteins known to have such features help form pumps and portals. The membranes of all living cells contain many different kinds of protein machines that act as gateways, allowing such molecules as foodstuffs or nutrients to pass in and waste products to pass out. Sometimes the gateways are passive, simply allowing the right-shaped molecules to float through on their own. Often, however, the portals are active, grabbing the right molecules and pushing them through the membrane. Because the cell has to deal with many different kinds of molecules that pass in both directions, it has many separate pumps and portals.

CHANGING THE PUMP

The amino acid sequence of PfCRT suggested that it was a protein pump, but that suspicion needed to be confirmed. Using clever laboratory techniques, Wellems and his coworkers demonstrated that the protein was located in the membrane of the parasite's digestive vacuole—its stomach. Thanks to this protein, the stomachs of mutant parasites accumulate a lot less chloroquine, and the bug survives to reproduce. Exactly why the mutant stomachs collect less drug is currently unclear, but it may well be that the mutation allows the chloroquine to leak out through the PfCRT pump.[6]

The staggering complexity of modern biology is a challenge for anyone to understand, but in order to find the edge of evolution, we need to get to the bottom of it—and we aren't quite there yet. The PfCRT protein has 424 amino acids. Just as sickle hemoglobin exhibits a change in its amino acid sequence from normal hemoglobin, the mutant PfCRT also has changes in its sequence.[7] And just as different hemoglobin mutants can be found in different areas of the world (such as HbC in Africa, HbE in Asia, and thalassemias around the Mediterranean Sea), different mutations have been found in PfCRT from different regions of the globe. Scientists have analyzed the protein from P. falciparum from patients in South America, Asia, and Africa. The mutant PfCRTs exhibit a range of changes, affecting as few as four amino acids to as many as eight. However, the same two amino acid changes are almost always present—one switch at position number 76 and another at position 220. The other mutations in the protein differ from each other, with one group of mutations common to chloroquine-resistant parasites from South America, and a second clustering of mutations appearing in malaria from Asia and Africa. This suggests that chloroquine resistance in malaria probably arose at least twice, separately in South America and Asia, and that the Asian resistance was transmitted to Africa. Later work suggested that there had actually been four separate origins.[8]

Since two particular amino acid changes occur in almost all of these cases, they both seem to be required for the primary activity

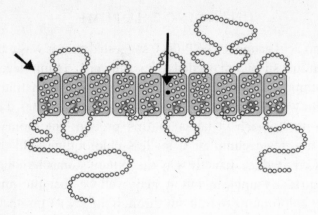

FIGURE 3.1
Schematic drawing of the PfCRT protein. Each circle represents an amino acid position. Arrows point to black circles that represent two positions (76 and 220) where mutations are almost always found in resistant proteins. (Reproduced from Bray, P. G., Martin, R. E., Tilley, L., Ward, S. A., Kirk, K., and Fidock, D. A. 2005. Defining the role of PfCRT in *Plasmodium falciparum* chloroquine resistance. *Mol. Microbiol.* 56:323–33. Courtesy of Blackwell Publishing.)

by which the protein confers resistance. The other mutations apparently "compensate" for side effects caused by these two primary mutations.

In the last chapter we saw that changes in human genes in the wake of malarial attacks were diminishments—beneficial only in dire circumstances, but detrimental in normal times. *P. falciparum*, however, greatly outnumbers humans, and reproduces much more rapidly, and therefore has many more opportunities for lucky genetic accidents. By standard Darwinian theory, it ought to make the next step in the arms race very early. Standard Darwinian logic predicts that malaria will mutate more, and sift its mutations more effectively, than humans. So are the changes in the mutated PfCRT an improvement? Is the parasite strengthening in an absolute sense, and evolving new "advanced and complex machinery," as Richard Dawkins might expect? It appears not. When chloroquine is no longer used to treat malaria patients in a region, the mutant strain of *P. falciparum* declines and the original strain makes a comeback, indicating that the mutant is weaker than the original strain in the absence of the toxic chloroquine.[9] Apparently, much like human thalassemia or sickle hemoglobin or G6PD deficiency, the mutant malarial protein is a net plus only in desperate circumstances—in trench warfare.

TIGER BY THE TAIL

As a teenager I was a big fan of science fiction, and I remember reading a short story entitled "Tiger by the Tail."[10] In the story an opening to another dimension somehow popped up in . . . a pocketbook! Someone had the bright idea to toss a grappling hook through the opening. If the grappling hook pulled stuff from the other dimension into ours, or vice versa, then the dimension that lost the tug of war would be destroyed. (There was a scientist on hand to explain it all.) This, the humans thought, was a great way to blackmail the aliens on the other side. As the story ended, however, the chain holding the grappling hook, which had been slowly emerging from the pocketbook, reversed direction as the aliens pulled harder on their end. The tables had turned, and now all humanity was threatened.

There is an analogy to the human-malaria struggle. The malarial parasite is turning, too, pulling harder on the grappling hook of synthetic drugs. In many ways chloroquine was a dream drug—not only effective but cheap, and with few side effects. It lasted for decades before resistance to it became widespread. Newer drugs and methods to combat malaria fall short on one or more of these features. Not only is malaria almost wholly resistant to chloroquine, but it is becoming increasingly adept at shrugging off the newer drugs that have followed. Malaria that has developed resistance to one drug seems to develop resistance to new drugs at an accelerated rate compared to "initial" malaria.[11] Appendix B details the specifics of several rounds of this trench war.

The development of drug resistance in malaria, like the development of the sickle cell gene and thalassemia in humans, is a crystal clear example of Darwinian evolution in action. We see it all right there—the selective pressure exerted on malaria by toxic drugs, the occasional mutations that make one bug more fit than its kin, the spreading of the mutation through the population. Yet malaria beautifully illustrates both the strengths and the shortcomings of the sort of blind search that Darwinian evolution demands. And with the help of mathematics, we can finally begin to achieve some precision about the limits of random mutation.

Answer: The obstacle that malaria hasn't been able to mutate
            around.
Question: What is sickle hemoglobin?

In our grudging admiration of *P. falciparum*'s wizardry at quickly mutating past our wonder drugs (even if the changes are ultimately diminishments), it's easy to lose sight of its failure to deal with sickle hemoglobin. Sickle has been around for thousands of years, not for mere decades like antimalarial drugs. Resistance to one recent drug, atovaquone, arose in the lab scant weeks after a small culture of malaria was exposed to it. Almost a hundred thousand times as many ticks of the clock have passed since Sickle Eve was born. About that much time has passed since C-Eve lived, too, and since thalassemia first appeared. Yet they are all still effective against malaria.

How can that be? Why should a single amino acid change in sickle hemoglobin checkmate malaria's million-murdering death when the best rational efforts of chemists are brushed aside in short order? Is the answer the complexity of the chemical? No. Chloroquine is no less complex than the new amino acid in sickle hemoglobin. Is it the specificity of the target? No. The problems chloroquine causes are no less specifically targeted to the stomach of the parasite than the problems sickle creates in the red blood cell; arguably chloroquine is more specific. Is it the inexperience of the parasite in dealing with the arena of attack? Hardly. The parasite routinely *eats* hemoglobin; it is *made* to deal with the stuff. Chloroquine is an artificial chemical. Malaria never saw the drug before the 1930s. Yet the parasite conquered chloroquine but is stumped by sickle.

I will get to the reason why sickle is such a challenge. At this point, let's simply take note of an exceptionally important implication of the widely varying success of malaria at dealing with the challenges that come its way. One simple yet crucial conclusion that we can already draw is this:

Darwinian evolution can deal quickly and easily with some problems, but slowly if at all with others.

Are there problems that are even harder for evolution than dealing with sickle hemoglobin? Problems that are for all intents and purposes beyond the reach of random mutation?

POWERBALL

Almost every Monday night my wife and I go to a sports bar a few miles from home to have a couple of hours alone together, away from the kids. (Buffalo wings are half price on Monday night.) On our way there I usually stop at a convenience store to buy a Powerball lottery ticket. It surely won't be long until I hit the jackpot, but so far I haven't managed to match more than a single number on any ticket. Luck of the Irish.

In the multistate Powerball lottery five white balls are drawn from a drum that contains fifty-three consecutively numbered balls, and then one (the fabled Powerball itself) from a separate container with forty-two red balls. The lottery ticket has five numbers listed together and then a separate number. There are various ways to win prizes. If the separate number on your ticket matches the Powerball, you get three dollars. If both the Powerball number and one other match, you're up to four dollars. To win the grand prize (at least ten million dollars) all the numbers have to match.

The odds of winning the grand prize are officially listed as one in 120,526,770. The odds of winning lesser prizes are better: one in 260 for matching three white balls, one in 12,248 for matching four white balls, one in 2,939,677 for matching five white balls. Not surprisingly, the odds of winning get worse the more balls that have to match, and they get worse rapidly. Matching four balls isn't just, say, 33 percent harder than matching three—it's about fifty times harder. Matching five balls is hundreds of times less likely than matching four, and ten thousand times less likely than matching three.

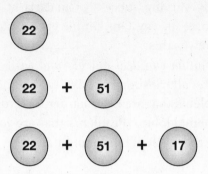

FIGURE 3.2
The odds against winning a lottery such as Powerball increase rapidly the more numbers that have to match. The same principle holds true for mutations to DNA.

Often in a Powerball drawing no one wins the jackpot, because no one bought a ticket that matched the winning numbers. When that happens the accumulated money is rolled over and the jackpot for the next drawing is that much bigger. How long will it likely take before someone (not a particular person—anyone at all) wins the grand prize? Besides the odds, the length of time depends on two other things—the number of players and the frequency of drawings. If the odds of winning are one in a hundred million, and if a million people play every time, then it will take on average about a hundred drawings for someone to win. So if there are two drawings per week (about a hundred per year), then it would take about a year before someone won. But if there were only one drawing per year, on average it would take a century to hit the jackpot. If the number of players were different, the average time to produce a winner would also change. If ten million people played each time instead of one million, then on average only ten drawings would be necessary to get a winner. If a billion people played each time, then it would be very likely that each drawing would have multiple winners.

The very same three simple considerations that regulate how often the Powerball lottery is won—the odds of winning, the number of players, and how often the lottery is held—also govern how fast malaria develops resistance to an antibiotic. Just as the odds of winning at Powerball depend on the range of possible numbers (1 to 53) and how many balls you have to match, the odds of developing antibiotic resistance depend on the number of nucleotides in the parasite's genome (millions) and how many mutations have to accumulate before there's a beneficial effect. Just as the time to get a winner depends on how frequently Powerball lotteries are held, the time to produce resistance in P. falciparum depends on the organism's mutation rate and generation span. And just as the time to a jackpot is shorter the more people play Powerball, the time to antibiotic resistance is shorter the more malarial parasites are exposed to a drug.

BACK STOP

Malaria scientists are acutely aware of these factors and take them into account when planning the battle against the parasite. One team of researchers notes that "the ease with which a population of parasites survives exposure to a drug depends on . . . the frequency at which the parasites develop resistance . . . and the size of the parasite population at the onset of treatment."[12] For example, if the chances that a cell will be resistant to a given drug are one in a million, and if there are a hundred million parasitic cells in a patient, then almost certainly there will be some resistant cells in the patient. Administering the drug will kill off the 99.9999 percent of cells that are sensitive, but the approximately one hundred resistant cells will survive and multiply. Soon the patient will be filled with resistant cells.

If the chances for resistance to a different drug are one in a billion, then odds are good that a person suffering from malaria who carries just a hundred million parasitic cells will be cured by the drug, which would likely kill all the cells. However, if ten people each have a hundred million cells, then altogether there are a billion P. falciparum cells, and there's an even chance that one of those cells in one of those patients will be resistant. If a clinic treats a hundred such patients a day, then on average one in every ten of those patients will harbor a resistant cell. When those patients are treated, the resistant cells will survive and replicate, the patients will be bitten by mosquitoes, and the mosquitoes will spread resistant cells to other people. In short order the drug will be useless.

To greatly increase the chances of successful treatment, one strategy is to use a cocktail of drugs, each component of which is able to kill a sizeable chunk of cells. For example, in urging that several drugs should be used simultaneously against malaria, one researcher explained:

> Resistance to antimalarial drugs arises when spontaneously occurring mutants . . . which confer reduced drug susceptibility are selected, and are then transmitted. Simultaneous use of two or more antimalarials . . . will reduce the chance of selection, because the chance of a resistant mutant surviving is the product of the parasite mutation rates for the individual drugs, multiplied by the number of parasites in an infection that are exposed to the drugs.[13]

Suppose a cocktail contains two drugs, A and B, and that one in a million parasite cells are resistant to drug A, and one in a million to drug B. Assuming resistance to A is due to a different mutation than resistance to B, then the odds that a single individual cell is resistant to both drugs at the same time are multiplied, a million times a million, which is one in a trillion.

In a real-world experiment involving this basic principle, scientists from Catholic University in Washington, D.C., showed that resistance to one drug (called 5-fluoroorotate) at a particular concentration was found in malaria cells at a frequency of about one in a million. Resistance to a second drug (atovaquone) was about one in a hundred thousand. Sure enough, resistance to both drugs was the multiplied odds for the two cases, about one in a hundred thousand times a million, that is, one in a hundred billion.

Using a combination of drugs is a common strategy to delay the onset of resistance. In addition to the battle with malaria, for example, drug cocktails are used in the fight against AIDS and tuberculosis. Delaying the onset of resistance, though, is not the same as stopping it altogether. The researchers from Catholic University warned that, although the combination of drugs they tested would be likely to cure any given person since the likelihood the person would harbor a resistant bug would be small, "a large enough patient population will inevitably allow selection of parasites that are resistant to both compounds." [14]

TWO FOR THE PRICE OF ONE

Suppose that *P. falciparum* needed several separate mutations just to deal with one antimalarial drug. Suppose that changing one amino acid wasn't enough. Suppose that two different amino acids had to be changed before a beneficial effect for the parasite showed up. In that case, we would have a situation very much like a combination-drug cocktail, but with just one drug. That is, the likelihood of a particular *P. falciparum* cell having the several necessary changes would be much, much less than the case where it needed to change only one amino acid. That factor seems to be the secret of why chloroquine was an effective drug for decades.

How much more difficult is it for malaria to develop resistance to chloroquine than to some other drugs? We can get a good handle on the answer by reversing the logic and counting up the number of malarial cells needed in order to find one that is immune to the drug. For instance, in the case of atovaquone, a clinical study showed that about one in a trillion cells had spontaneous resistance. [15] In another experiment it was shown that a single amino acid mutation, causing a change at position number 268 in a single protein, was enough to make *P. falciparum* resistant to the drug. So we can deduce that the odds of getting that single mutation are roughly one in a trillion. On the other hand, resistance to chloroquine has appeared fewer than ten times in the whole world in the past half century. Nicholas White of Mahidol University in Thailand points out that if you multiply the number of parasites in a person who is very ill with malaria times the number of people who get malaria per year times the number of years since the introduction of chloroquine, then you can estimate that the odds of a parasite developing resistance to chloroquine is roughly one in a hundred billion billion. [16] In shorthand scientific notation, that's one in 10^{20}.

BOX 3.1
Scientific Notation

Scientists often have to deal with numbers that are very large (say, the number of stars in the universe) or very small (say, the mass of a proton). To do so conveniently, scientific notation can be used. In scientific notation a ten is written and the number of zeroes in the number is written as a superscript to the right of the ten. For example, instead of 10, 100, and 1,000, the numbers ten, one hundred, and one thousand are written as 10^1, 10^2, and 10^3, respectively. Instead of writing out a big number such as a trillion as 1,000,000,000,000, in scientific notation a trillion is written simply as 10^{12}, which is easier on the eyes and saves space. One has to keep in mind that numbers increase very, very quickly as the superscript (called the exponent) increases. For example, compare the numbers 10^4 and 10^{10}. They might not seem so different at first blush. However, a minimum wage worker might earn 10^4 (ten thousand) dollars per year; only someone like Bill Gates might earn 10^{10} (ten billion) dollars per year. The difference of just six between the ten and the four in the exponents means that the numbers differ by a million-fold. The figure below shows a scale with numbers written in both common and scientific notation, and corresponding population numbers to put things in perspective.

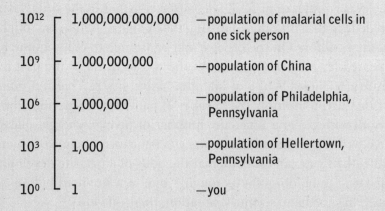

10^{12}	1,000,000,000,000	—population of malarial cells in one sick person
10^9	1,000,000,000	—population of China
10^6	1,000,000	—population of Philadelphia, Pennsylvania
10^3	1,000	—population of Hellertown, Pennsylvania
10^0	1	—you

MATCHING TWO

Let's compare the two numbers for the odds of achieving resistance to atovaquone, where just one mutation is needed, versus chloroquine, where (presumably—since if a single mutation could help, chloroquine resistance would originate much more frequently) two are needed. The odds are, respectively, one in a trillion (10^{12}) and one in a hundred billion billion (10^{20}). The ratio of the two numbers shows that the malarial parasite is a hundred million times (10^8) less likely to develop resistance to chloroquine than to atovaquone. This is reasonable since the genome size of the malarial parasite is in the neighborhood of a hundred million nucleotides. The implication is that if two amino acids in a protein have to be changed instead of just one, that decreases the likelihood of resistance by a factor of about a hundred million.

Even though the odds are tremendously stacked against it, *P. falciparum* was able to develop chloroquine resistance because there are an enormous number of parasitic cells (about a trillion) in an infected patient's body, and about a billion infected people in the world in a year. So the parasite has the population numbers to get around the terrible odds. Spontaneous resistance to atovaquone can be found in roughly every third sick person.[17] Spontaneous resistance to chloroquine can be found perhaps in every billionth sick person, and since there are usually close to a billion sick people on the planet every year or so, that means chloroquine resistance is usually waiting to be found in at least one person, somewhere in the world, at any given time.

FEWER PLAYERS, LONGER TIMES

Suppose that *P. falciparum* were not quite as prodigious as it actually is. What if, instead of a trillion malarial cells in the typical sick person, there were only a million? How long would it then take for chloroquine resistance to pop up? If all other things were equal, it would take about a million years. The reason is that if there were fewer parasites per person, and therefore fewer in the world's population, then the parasite would have to wait a proportionately longer amount of time for the right combination of mutations to come along. The number of players in the lottery would be decreased a millionfold, so the length of time needed to get a winner would be increased a millionfold.

This straightforward example carries an obvious implication. Species in which there are fewer living organisms than malaria (again, other things being equal) will take proportionately longer to develop a cluster of mutations of the complexity of malaria's resistance to chloroquine. Let's dub mutation clusters of that degree of complexity—1 in 10^{20}—"chloroquine-complexity clusters," or CCCs. Obviously, since malaria is a microbe, its population is far more vast than any species of animal or plant we can see with the unaided eye. Virtually any nonmicroscopic species would take longer—perhaps much, much longer—to develop a CCC than the few years in which malaria managed it, or the few decades it took for that mutation to spread widely.

Consider a species that is dear to our hearts—*Homo sapiens*. The number of human players in the world is much fewer than 10^{20}. For most of the past ten million years the population of the line of primates leading to humans is thought at best to have been roughly about a million or so.[18] Only in the past few thousand years did that number accelerate up to today's population of 6 billion.

What is the total number of creatures in the line leading to humans since it split from the line leading to modern chimps less than ten million years ago? If the average generation span of humanoids is rounded down, conservatively, to about ten years, then a generous estimate is that perhaps a trillion creatures have preceded us in the past ten million years.[19] Although that's a lot, it's still much, much less than the number of malarial parasites it takes to develop chloroquine resistance. The ratio of humanoid creatures in the past ten million years to the number of parasites needed for chloroquine resistance is one to a hundred million.

If all of these huge numbers make your head spin, think of it this way. The likelihood that *Homo sapiens* achieved any single mutation of the kind required for malaria to become resistant to chloroquine—not the easiest mutation, to be sure, but still only a shift of two amino acids—the likelihood that such a mutation could arise just once in the *entire* course of the human lineage in the past ten million years, is minuscule—of the same order as, say, the likelihood of you personally winning the Powerball lottery by buying a single ticket.

On average, for humans to achieve a mutation like this by chance, we would need to wait a hundred million times ten million years. Since that is many times the age of the universe, it's reasonable to conclude the following: *No mutation that is of the same complexity as chloroquine resistance in malaria arose by Darwinian evolution in the line leading to humans in the past ten million years.*

Instead of concentrating on us humans, we can look at the odds another way. There are about five thousand species of modern mammals. If each species had an average of a million members,[20] and if a new generation appeared each year, and if this went on for two hundred million years,[21] the likelihood of a single CCC appearing in the whole bunch over that entire time would be only about one in a hundred.

Let that sink in for a minute. Mammals are thought to have arisen from reptiles and then diversified into a spectacular array of creatures, including bats, whales, kangaroos, and elephants. Yet that entire process would—if it occurred through Darwinian mechanisms—be expected to occur without benefit of a single mutation of the complexity of a CCC. Strict Darwinism requires a person to believe that mammalian evolution could occur without any mutation of the complexity of this one.

Here's a possible point of confusion. We estimated the odds of a CCC—one in a hundred billion billion (10^{20})—by looking at the number of malarial parasites needed to develop the double mutation of a particular protein of a particular gene. Someone might object that, since there are thousands of other proteins in an organism, much other DNA, and many other kinds of mutations than just amino acid changes, aren't the odds of finding *some* beneficial complex of mutations much better than the odds of finding just the specific complex of mutations we isolated?

No. Many, many other mutations in addition to the ones we discussed have popped up by chance in the vast worldwide malarial pool over the course of a few years. In fact mutations in *all* of the amino acid positions of *all* of the proteins of malaria—taken both one and two at a time—can be expected to occur by chance during the same stretch of time. And other types of mutations besides just changes in amino acids would also occur (such as insertions, deletions, inversions, gene duplications, mobile DNA transpositions, changes in regulatory regions, and others, perhaps even including whole genome duplication—some of these types of mutations are discussed in the next chapter). Although some other mutations in some other proteins are thought to contribute to chloroquine resistance,[22] none are nearly as effective as that in PfCRT. That means that of all of the possible mutations in all of the different proteins of malaria, only a minuscule number have the ability to help at all against chloroquine, and only one, PfCRT, is really effective. Natural selection gets to choose from a staggering number of variations, yet at best only a handful help. So a CCC isn't just the odds of a particular protein getting the right mutations; it's the probability of an effective cluster of mutations arising in an entire organism.

EVEN WORSE

The development of chloroquine resistance isn't the toughest problem that evolution faces. We know that for certain, because the malarial parasite solved that problem but hasn't solved others, such as sickle hemoglobin. How much more difficult than a CCC would a challenge have to get before Darwinian evolution would essentially be ineffective, even for simple single-celled creatures such as malaria?

First think of it this way. What if, to win a super-Powerball lottery, instead of matching all the numbers on *one* ticket, some person had to match all the numbers on *two* tickets? The likelihood of that happening would be about the square of the odds of matching the numbers on one ticket, roughly a hundred million squared. If that were the case, then (if other things were equal) it would take millions of years for any person at all to win the lottery.

Recall that the odds against getting two necessary, independent mutations are the multiplied odds for getting each mutation individually. What if a problem arose during the course of life on earth that required a cluster of mutations that was twice as complex as a CCC? (Let's call it a double CCC.) For example, what if instead of the several amino acid changes needed for chloroquine resistance in malaria, twice that number were needed? In that case the odds would be that for a CCC times itself. Instead of 10^{20} cells to solve the evolutionary problem, we would need 10^{40} cells.

Workers at the University of Georgia have estimated that about a billion billion trillion (10^{30}) bacterial cells are formed on the earth each and every year.[23] (Bacteria are by far the most numerous type of organisms on earth.) If that number has been the same over the entire several-billion-year history of the world, then throughout the course of history there would have been slightly *fewer* than 10^{40} cells, a bit less than we'd expect to need to get a double CCC. The conclusion, then, is that the odds are slightly against even one double CCC showing up by Darwinian processes in the entire course of life on earth.

Put more pointedly, a double CCC is a reasonable first place to draw a tentative line marking the edge of evolution for all of life on earth. We would not expect such an event to happen in all of the organisms that have ever lived over the entire history of life on this planet. So if we do find features of life that would have required a double CCC or more, then we can infer that they likely did not arise by a Darwinian process.

As we'll see, life is bursting with such features.

MAKING DISTINCTIONS

We've come a long way in a short space by drawing out implications from the long trench war of attrition between humanity and malaria. Perhaps, however, we've moved a bit too fast. Even with its limited resources, Darwinian evolution has a number of tricks up its sleeve, tricks that can easily be overlooked if you're not careful. In order to be as confident as we can of where to draw the line marking the edge of Darwinian evolution, we need to have a thorough appreciation for what random mutation can do. In the next chapter we'll survey the kinds of tools that are available to evolution and look at examples of where it has acted.

4

WHAT DARWINISM CAN DO

COMMON DESCENT VERSUS RANDOM MUTATION

"How stupid of me not to have thought of it!" So lamented the naturalist Thomas Huxley upon first hearing of Darwin's theory of evolution. While his ideas may not explain all of biology, from the moment they were published in 1859 all biologists have realized that they do explain a great deal. In this chapter we'll focus on what clearly can be explained by Darwin.

Bear in mind, throughout, that common descent is a distinct concept from the mechanism of natural selection acting on random mutation. It isn't always easy to keep them apart. In practice, if you're not careful, it's easy to mistake the effects of common descent for the effects of natural selection. In fact, it's so easy that even Darwin himself mixed them up. Writes Ernst Mayr:

> That writers on Darwin have nevertheless almost invariably spoken of the combination of these various theories as "Darwin's theory" in the singular is in part Darwin's own doing. He not only referred to the theory of evolution by common descent as "my theory," but he also called the theory of evolution by natural selection "my theory," as if common descent and natural selection were a

single theory. . . . [Darwin] ascribed many phenomena, particularly those of geographic distribution, to natural selection when they were really the consequences of common descent.[1]

To find the edge of evolution we need to take care to distinguish the two. Although human-malaria trench warfare shows that random mutation is severely limited in scope, the idea of common descent has a lot more going for it.

Descent is often the aspect of Darwin's multifaceted theory that is most emphasized. For example, in the final sentence of *The Origin of Species* Darwin waxed lyrical.

There is grandeur in this view of life, with its several powers, having been originally breathed by the Creator into a few forms or into one; and that, whilst this planet has gone cycling on according to the fixed law of gravity, from so simple a beginning endless forms most beautiful and most wonderful have been, and are being evolved.[2]

Over the next few sections I'll show some of the newest evidence from studies of DNA that convinces most scientists, including myself, that one leg of Darwin's theory—common descent—is correct. Let's begin by looking at something Darwin knew nothing about—the genetic basis of life, and how it can change.

VARIETY SHOW

In *The Origin of Species* Darwin proposed that natural selection acts on variation in the living world, rewarding the more fit and weeding out the less fit. At the time the underlying basis for variation within a species was unknown. Darwin had to simply assume that there was some mechanism, unknown to the science of his age, to generate differences.

One of the greatest triumphs of twentieth-century science was its discovery of the basis of biological inheritance. In a classic experiment in the 1940s Oswald Avery showed that DNA is the carrier of genetic information. Watson and Crick deciphered the elegant double helical shape of that molecule. Marshall Nirenberg cracked its genetic code. More recently, scientists developed methods to clone, synthesize, and sequence DNA. In June 2000 President Clinton and Great Britain's prime minister Tony Blair jointly announced the completion of the sequencing of the human genome. The announcement marked an unparalleled milestone in human intellectual achievement. Yet it was only a way station, not a terminal, in the investigation into the foundation of life on earth. Since then the genomes of hundreds of other organisms have been sequenced, and thousands more are planned. Most of those organisms are single-celled microbes, whose genomes are much smaller (about one-thousandth the size) than those of animals like us. But the genomes of some larger plants and animals have also been sequenced, including those of the chimp, dog, zebrafish, and rice.

Rapidly accumulating data from genome sequencing projects have allowed scientists to look at the many different ways DNA can change. In other words, only in recent decades have we been able to examine the kinds of variations—mutations—that can spring up in a genome. The cellular machinery that replicates DNA is extremely faithful. In people and other multicellular organisms it makes only about one mistake in every hundred million nucleotides of DNA it copies in a generation. Yet since the number of nucleotides in a cell's genome is on the order of millions to billions, on a per-cell basis, mistakes actually happen pretty often. On average, depending on the kind of organism and how much DNA it has, a mutation happens at a rate from about once every hundred cells to ten mutations per cell. If DNA were exactly like a blueprint, with no wasted space, and every line and curve representing a point of building, then this mutation rate would be fatal. After all, one critical mistake is all it takes to kill (or cause the building to collapse). But in fact, DNA isn't exactly like a blueprint. Only a fraction of its sections are directly involved in creating proteins and building life. Most of it seems to be excess DNA, where mutations can occur harmlessly.

Mutations come in different flavors. When Sickle Eve was conceived, one copy of the DNA section that served as a blueprint for the beta chain of her hemoglobin was altered, so that a single amino acid was substituted for another. That, unsurprisingly, is called a substitution mutation—a straightforward switch, where one single letter of the billions in DNA is traded for another. A single-letter substitution often leads to a change in a protein amino acid sequence, as it did with Sickle Eve, but not always.[3]

There are other kinds of mutations, too. One class is called deletion mutations. As the name implies, deletion mutations occur when a portion of DNA, ranging from a single letter to a large chunk of the genome, is accidentally left out when the DNA is duplicated. For example, some people have thirty-two nucleotides (letters) deleted in a gene for a protein called CCR5. Blessedly, the mutant gene confers resistance to HIV, the virus that causes AIDS in humans. The opposite of a deletion mutation is an insertion mutation. This happens when extra DNA is accidentally placed into a region. People who suffer from Huntington's disease (such as 1930s folk singer Woody Guthrie) have many extra copies of a particular three-nucleotide segment (C-A-G) in the gene for a protein called huntingtin.[4] Like deletions, insertions can range from just one letter to many. Sometimes the insertion happens because the molecular machinery copying DNA "stutters," backs up, and recopies a region it has just copied, so that a piece of DNA is copied twice. Other times a large piece of DNA (thousands of nucleotides in length) from an active element from one region of DNA copies itself into the genome where it hadn't been before.

TABLE 4.1
Varieties of DNA Mutations

Type of Mutation	Description
substitution	switch of one kind of nucleotide for another
deletion	omission of one or more nucleotides
insertion	addition of one or more nucleotides
inversion	"flipping" of a segment of DNA double helix
gene duplication	doubling of a region of DNA containing a gene
genome duplication	doubling of the total DNA of an organism

A special kind of insertion occurs when the extra DNA comes from a different organism. Viruses are small scraps of genetic material, either DNA or RNA, that invade cells and use the cells' resources to copy themselves. Sometimes they insert their own DNA into the host genome, where it can remain indefinitely. Other times, while a virus is replicating its own genome, a piece of the cell's genetic material accidentally gets picked up and added to that of the virus. If the extra material does the virus more good than harm, it can become a permanent part of the viral genome.

Another kind of mutation is called an inversion. When some of the normal machinery of the cell goes slightly awry, a piece from the DNA double helix can be cut out, flipped over, and stitched back in. This sort of mutation is thought to help divide one species into two species. Organisms with inverted regions in their DNA can mate with each other, but they often cannot mate as successfully with their "unflipped" cousins. One species of mosquito that carries malaria in west Africa seems to be dividing into several separate species because of large genomic inversions.[5]

Another type of mutation, thought by Darwinists to be especially consequential, is gene duplication. Occasionally an entire gene or set of genes gets copied twice on a chromosome, so the mutant organism now has two or more copies of a gene where its kin have only one. For example, laboratory resistance to chloroquine was seen in some malarial cells that mutated extra copies of segments of the parasite's chromosome 3.[6] When genes accidentally duplicate, evolution has a golden opportunity. Now one copy of the gene can continue to take care of its original job, while the second, spare copy of the gene is free to be used for a different job. We'll see later on that, although gene duplication can help in limited circumstances, like Darwinian processes in general it doesn't take us very far.

How often do mutations occur? Any one particular nucleotide (like, say, the one that will give the sickle mutation) is freshly substituted about once every hundred million births.[7] Small insertion and deletion mutations pop up roughly at the same rate. Gene duplications also seem to occur at about the same frequency.[8] So if the population size of a species is a hundred million, then on average *each and every* nucleotide is substituted in some youngster in *each* generation, and *each* gene is also duplicated in someone, somewhere. And so on. On the other hand, if the population size is only a hundred thousand, it would take a thousand generations for a duplicate of a particular gene or a particular nucleotide substitution to arise (on average)—because that's how long it would take to reproduce a hundred million organisms.

A word of caution. Although substitutions, insertions, deletions, and duplications all happen roughly at the same rate, there is a critical distinction between breaking something old and building something new. It's always easier and faster to blow up a bridge than to build one. For example, in human history a new sickle mutation (not one that was just inherited from a parent who had it) has freshly arisen at most only a few times, perhaps just once. Yet thalassemia has popped up hundreds of times. The reason for the difference in the numbers is easy to see. To get sickle, one particular nucleotide has to be substituted. To get thalassemia (which breaks a hemoglobin gene), on the other hand, any of hundreds of nucleotides can be substituted or deleted and the gene will no longer produce a working protein. Any of a large number of substitutions or deletions will suffice. In general, then, mutations that help in trench warfare by breaking something will appear at a rate hundreds of times faster than ones that help by doing something new.

YOU CAN PICK YOUR FRIENDS, BUT . . .

Scientific work in earlier centuries first noted the remarkable anatomical similarities between humans and other primate species. With the advent of modern biology, the sequences of their protein and DNA could also be compared.

One of the side benefits of our new understanding of DNA is that scientists can often use it to figure out who is related to whom. For example, DNA tests can establish paternity in disputed cases, or determine which side of the family a genetic disease has come from. This can infer relationships not only among modern humans, but with ancient ones, too. By comparing protein and DNA sequences, the origin of Sickle Eve can be pinpointed with reasonable accuracy. In the 1980s scientists compared data from modern humans and proposed the hypothesis of "Mitochondrial Eve"—that all modern humans are descended from a single woman who lived perhaps a hundred thousand years ago.

Although it is trickier and depends on more assumptions, the same general sorts of methods and reasoning that establish relationships among modern humans, and between modern and ancient humans, are also used to figure out how different species are related to each other. If two kinds of organisms share what seems to be a common mutation or set of mutations in their DNA, it can be assumed that a common ancestor of the two species originally suffered the mutation, and the descendants simply inherited it. Admittedly, assumptions are involved, but they strike many people as reasonable.

In the early 1960s the first sequences of proteins became available. Scientists were shocked. Many had expected the biological molecules of different organisms to be completely different. But the molecules often turned out to be similar in a very suggestive way. For example, one of the first proteins to be sequenced from a wide variety of organisms was hemoglobin. The sequence of hemoglobin in various species reflected the biological classification system that had been set up centuries earlier. The amino acid sequence of the beta chain of human hemoglobin was much different from that of fish, somewhat different from that of kangaroo (a marsupial mammal), pretty similar to that of dog (a placental mammal), and *identical* to that of chimpanzee.[9] The protein pattern fit wonderfully with Darwin's image of a branching tree of life. Not only hemoglobin, but many other molecular similarities were discovered between humans and other primates and, more broadly, underlying all of life.

One serious objection might be raised. Perhaps the different animals all had similar hemoglobin because that's the only protein that could really work to carry oxygen efficiently. Just as all organisms have to be based on carbon, because carbon is the only element versatile enough for life, perhaps all animals simply have to have certain similarities in their molecular machinery. So by necessity any large animal would have to have a protein similar to hemoglobin, even if it arose separately.

That objection, however, doesn't hold for a feature shared between two organisms that has no functional role to play. When two lineages share what appears to be an arbitrary genetic accident, the case for common descent becomes compelling, just as the case for plagiarism becomes overpowering when one writer makes the same unusual misspellings of another, within a copy of the same words. That sort of evidence is seen in the genomes of humans and chimpanzees. For example, both humans and chimps have a broken copy of a gene that in other mammals helps make vitamin C. As a result, neither humans nor chimps can make their own vitamin C. If an ancestor of the two species originally sustained the mutation and then passed it to both descendant species, that would neatly explain the situation.

More compelling evidence for the shared ancestry of humans and other primates comes from their hemoglobin—not just their working hemoglobin, but a broken hemoglobin gene, too.[10] In one region of our genomes humans have five genes for proteins that act at various stages of development (from embryo through adult) as the second (betalike) chain of hemoglobin. This includes the gene for the beta chain itself, two almost identical copies of a gamma chain (which occurs in fetal hemoglobin), and several others. Chimpanzees have the very same genes in the very same order. In the region between the two gamma genes and a gene that works after birth, human DNA contains a broken gene (called a "pseudogene") that closely resembles a working gene for a beta chain, but has features in its sequence that preclude it from coding successfully for a protein.

Chimp DNA has a very similar pseudogene at the same position. The beginning of the human pseudogene has two particular changes in two nucleotide letters that seem to deactivate the gene. The chimp pseudogene has the exact same changes. A bit further down in the human pseudogene is a deletion mutation, where one particular letter is missing. For technical reasons, the deletion irrevocably messes up the gene's coding. The very same letter is missing in the chimp gene. Toward the end of the human pseudogene another letter is missing. The chimp pseudogene is missing it, too.

The same mistakes in the same gene in the same positions of both human and chimp DNA. If a common ancestor first sustained the mutational mistakes and subsequently gave rise to those two

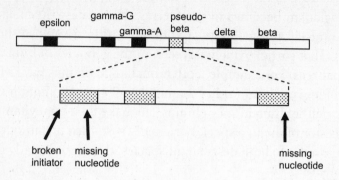

FIGURE 4.1

Human and chimp hemoglobin genes are very similar. The top bar is a schematic illustration of the region of the primate genomes that contain genes for the betalike chains of hemoglobin, including the pseudo-beta gene (in gray), which cannot produce a functional protein. The arrangement is identical for both humans and chimps. The bottom bar is an expanded view of the pseudo-beta gene. Gray regions correspond to regions of functional genes that code for part of the protein. Both human and chimp pseudo-beta genes contain the same mistakes that preclude making a working protein.

modern species, that would very readily account for why both species have them now. It's hard to imagine how there could be stronger evidence for common ancestry of chimps and humans.

That strong evidence from the pseudogene points well beyond the ancestry of humans. Despite some remaining puzzles,[11] there's no reason to doubt that Darwin had this point right, that all creatures on earth are biological relatives.

The bottom line is this. Common descent is true; yet the explanation of common descent—even the common descent of humans and chimps—although fascinating, is in a profound sense *trivial*. It says merely that commonalities were there from the start, present in a common ancestor. It does not even begin to explain where those commonalities came from, or how humans subsequently acquired remarkable differences. *Something that is nonrandom must account for the common descent of life.*

HE'S TWICE THE FUNGUS HIS DADDY WAS

The work on the hemoglobin genes of humans and chimps was done several decades ago. More recent work on whole genomes of yeast species further shows the power of the idea of common descent. Even better, this line of analysis has produced some of those eureka moments that make science so exciting—moments when newly accessible data suddenly illuminate a murky landscape like a flare in the night. It also points to the limits of random mutation.

Although most people think of yeast as the active agent that leavens bread or gives beer its zip, biologists classify yeasts as fungi—distant relatives of animals and plants. Scientists who work on yeast had long been suspicious of some features of the DNA of baker's yeast (whose scientific name is *Saccharomyces cerevisiae*).[12] It contains a number of genes that code for very similar proteins that seem to have almost redundant roles in the cell. The odd arrangement of genes led a couple of groups of scientists to hypothesize that, sometime in the misty past, perhaps a baker's yeast cell was born with the mother of all gene duplications. Instead of just one gene, or a chunk of the genome, the entire DNA of the yeast was duplicated! Instead of the roughly 12 million nucleotides that its brothers and sisters had, the prodigy had 24 million. At one stroke the offspring was literally twice the fungus his daddy was. Over time, however, much of the duplicated DNA was lost by deletion mutations.

That was the hypothesis—but how to test it? With just the sequence of baker's yeast (*S. cerevisiae*) DNA to go on, the suspicions couldn't be confirmed. So a French group sequenced the entire genomes—tens of millions of nucleotides—of four other diverse kinds of yeasts. The researchers saw that duplicate genes in baker's yeast could be lined up with their counterparts in the other yeasts. When they were aligned, one copy of a duplicated baker's yeast gene would sometimes be next to the left half of some genes that formed a single group in another yeast species, while the second baker's yeast gene copy would be next to the right half of the group in a separate region of the baker's yeast genome. That arrangement is consistent with the hypothesis—made years before

the genomes were sequenced—that the whole yeast genome duplicated and then many duplicate genes were deleted over time.[13]

This is yet more evidence for common descent. On the other
hand, the genome duplication seems not to have done a whole lot
for its recipients. All five yeasts have similar cell shapes and
lifestyles.[14] The duplicated baker's yeast has the ability to make
alcohol, but one unduplicated yeast can eat petroleum, arguably a
trickier business. Another yeast species, containing more duplicated DNA than baker's yeast, avoided whole-genome duplication;
it apparently duplicated genes the old-fashioned way—one by one
(or in blocks). Darwinists like to think that genome duplication is
one of the magic bullets of random mutation—it suddenly granted
vast new possibilities to the genome. Yet genome duplication—a
spare copy of *each and every gene* to play with—and a hundred million years of time seem not to have given baker's yeast any advantage it wouldn't otherwise have had.[15] This leads to a very important
point. Randomly duplicating a single gene, or even the entire
genome, does not yield new complex machinery; it only gives a
copy of what was already present. Although duplicated genes can be
used to trace common ancestry, neither individual gene duplications nor whole genome duplications by themselves explain novel,
complex forms of life.

INCH BY INCH

If genetics has supported common descent, what of the usefulness
of random mutation? It has fared decidedly less well, but still has
some victories to boast of. Darwin argued that evolution had to
work by tiny, random, incremental changes that improved the likelihood that a mutant organism would survive and prosper. So whenever we see such small beneficial changes or series of such changes,
we should tip our hat to the sage of Down House. Sickle Eve was
one example, as were the mutations that confer chloroquine resistance on malaria. To drive home the point that Darwinian random
mutation can certainly explain some simple features of life, in the
rest of the chapter I'll recount several more cases, beginning with a
few malaria-related examples.

As malaria developed resistance to the wonder drug chloroquine, scientists rushed to develop new treatments. One successor drug is called pyrimethamine. Interestingly, malaria can counter it with a single amino acid substitution. That single amino acid change makes malaria one hundred times more resistant to the drug. Malarial DNA has only about 23 million nucleotides. A sick person can be burdened with as many as a trillion parasite cells. If you do the math, the resistance mutation should occur by chance in at least one parasitic cell in almost every sick person. Looked at another way, resistance should develop independently many times over in a large group of patients treated with the drug. But a recent report by scientists at the National Institutes of Health pointed out a conundrum.

> Because resistance to [pyrimethamine] can be conferred by a single point mutation, it was assumed that resistance could occur frequently. However, a recent population survey demonstrated a single origin of [resistant genes] in five countries: Thailand, Myanmar, PDR Lao, Cambodia, and Vietnam.[16]

In other words, even though initial resistance springs up quickly and easily, and therefore mutant genes from many different malarial cells might be expected to be present in a country, only one gene from one original cell dominates a region up to a thousand miles across. How could that be?

Although the first mutation (at position 108 of the protein, as it happens) grants some resistance to the drug, the malaria is still vulnerable to larger doses. Adding more mutations (at positions 51, 59, and a few others) can increase the level of resistance. However, as usual there's a hitch. Some of those extra mutations (but not the first one) seem to interfere with the normal work of the protein. Perhaps, though, if other mutations in other genes could compensate for these harmful effects, greater resistance could be acquired without causing harm in the process. In other words, to move to the next level of resistance after the first mutation, two further, simultaneous mutations seem to be necessary. As the scientists point out, "Because concurrent mutations in two different genes occur at reduced frequency, this would help explain the rarity with which resistance has evolved."[17] Nonetheless, because malaria grows to huge population numbers—numbers that are much greater than those of mammals or other vertebrates—it can overcome poor odds. Apparently, as for the case of chloroquine resistance, a very lucky malarial cell in one infected person acquired the several changes that gave it greater resistance to pyrimethamine while compensating for any bad side effects. That rare mutant then spread quickly through the population. That double mutant is, it seems, roughly as rare as a CCC.

A second example of what natural selection can do comes from the poor, hijacked mosquito, which involuntarily carries malaria from human to human. In 1946 the insecticide DDT was first turned against the mosquito in order to fight the disease. Taking a page from Sickle Eve's book, mutant mosquitoes resistant to the chemical first showed up promptly in 1947. Mosquitos can resist DDT if they have mutations in their genes for enzymes whose normal job is to detoxify chemicals.

So, in the wake of the failure of DDT to control mosquitoes, other insecticides have been developed. One kind of insecticide targets an enzyme that is needed for the insect's nervous system to work. Although the chemical had previously been used on flies, which eventually developed resistance, it hadn't been widely used on mosquitoes, and no resistant mosquitoes had yet been discovered. To see if mosquitoes might develop resistance, some researchers deliberately altered the mosquito gene in the lab with the same mutations that made flies resistant. Sure enough, the altered mosquito gene became resistant, too. What's more, the workers showed that only one amino acid change was needed to achieve resistance, and that adding other mutations in the right places could increase that resistance.[18]

Although it hasn't yet occurred in nature, we shouldn't be at all surprised to see resistance of mosquitoes to the new insecticides arise and spread by Darwinian processes. The necessary preconditions are all there: tiny, incremental steps—amino acid by amino acid—leading from one biological level to another.

There's another very important lesson to be drawn from the fly/mosquito reaction to insecticides, a lesson pointing strongly to the limitations on Darwinian evolution. Mutation has to work with the pre-existing cellular machinery, so there is a very limited number of things it can do.[19] Even though there are trillions upon trillions of possible simple mutations to an insect's genome, *all but a handful are irrelevant*. The same few mutations pop up in organisms as disparate as mosquito and fly because no others work.

This limitation compounds the limitation noted earlier, that most mutations decrease an organism's overall functioning—they are destructive, not constructive, even among the tiny fraction of mutations that "work." Consider the example of the rat poison known as warfarin. It was developed in the 1950s. Warfarin interferes with the function of the blood-clotting system of mammals, so that a rat who eats it bleeds to death. Soon after warfarin was introduced, it lost effectiveness. It turns out that a change of any one of several amino acids in a certain rat protein is enough to confer resistance.[20] The likelihood of one of those particular amino acids mutating is on the order of a paltry one in a hundred million. However, since there are probably at least ten times that many rats in the world, the odds of some rat somewhere having the alteration are actually very good. In fact, the resistance mutation has arisen independently about seven times in the same protein.[21]

Looked at a different way, however, warfarin resistance points not to the strength of random mutation, but to its limitations. Since the same mutation has been selected a number of times, even though the worldwide population of rats contains much variation in all rat proteins, this strongly suggests that the only effective mutations are ones to that single protein. What's more, although they confer resistance to warfarin, the mutations also decrease the effectiveness of the enzyme, so it only works about half as well as the normal protein. In other words, as with many other mutations we've seen, the change is a net benefit only in desperate times.

FROZEN FISH

The examples of Darwinian natural selection discussed so far have all been relatively recent. Resistance to modern pesticides such as rat poison and chloroquine developed in just the past few decades.

Even the mutations that first led to Sickle Eve and thalassemia occurred no more than ten thousand years ago. There are two reasons for concentrating on relatively recent examples: First, our information about them is pretty solid, and much less tainted by the flights of imagination that plague most Darwinian storytelling; and second, the recent examples are widely touted by fans of Darwin as our best examples of natural selection in action. My final example of what Darwin can do, however, is much older, and so is a lot fuzzier. We can't easily determine the steps along the older example's pathway or measure the advantage of each in a laboratory. Nonetheless, it seems reasonably convincing.

Over ten million years ago currents in the waters around Antarctica began to form a closed loop, circling around and around the southernmost continent. With no warmer water from other parts of the globe flowing through, the temperature of the Antarctic Ocean slowly decreased until ice formed. Because the ocean contains salt, which lowers the freezing point of water, the temperature of the liquid sea decreased below the freezing point of pure water, and then decreased below the freezing point of bodily fluids. Since fish are cold-blooded animals whose body temperature is the same as that of the water they swim in, they were in danger of freezing solid as the environment changed.

Fast forward ten million years. One group of fish, called notothenioids, flourish in the Antarctic ocean, even though the ocean temperature is below the freezing point of their blood. How can they apparently defy the laws of physics? Why aren't they naturally frozen filets by now?

Notothenioids can flout the ice because they make some amazing proteins that literally stop water from freezing. When pure water is cooled below the freezing point it doesn't solidify right away. That's because a large number of water molecules first have to stick together to form tiny seed crystals. Once formed, the tiny crystals rapidly grow larger until all of the water has solidified. But if no seed crystals form, the water can stay liquid indefinitely, even below the freezing point. To make a long story short, antifreeze proteins stick to ice crystal seeds and stop them from growing. No seeds, no ice growth.

In 1997 a group of scientists at the University of Illinois sequenced the gene for an antifreeze protein from Antarctic fish. They were startled to discover so-called control regions to the left and the right of the portion of the gene that coded for the antifreeze protein that were very similar to control regions for another protein, a digestive enzyme.[22] Both portions had a certain nine-letter sequence, but in the antifreeze gene the nine-nucleotide region was repeated many times. This gave the protein a simple sequence that consisted of three amino acids repeated many times over.[23]

The scientists proposed that the antifreeze protein evolved in a Darwinian fashion, by random mutations and natural selection, beginning with a duplicate copy of the digestive-enzyme gene. A probable scenario goes something like the following: The first copy of that gene simply continued its normal job. But by chance, in one of the fish in the ancient Antarctic regions, the cell's machinery stuttered

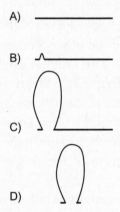

FIGURE 4.2
Schematic illustration of a possible evolutionary pathway of a simple antifreeze protein by small random mutations. A) A second copy of a digestive protein gene is produced by gene duplication. B) A nine-nucleotide region of the gene is accidentally duplicated (the small bump on the line), yielding a simple three amino acid repeat in the protein that has some antifreeze activity. C) "Stuttering" during DNA replication produces many more copies of the simple nine-nucleotide repeat (the enlarged bump), improving antifreeze properties of the protein. D) Regions of the gene that don't contribute to antifreeze activity are accidentally deleted.

when copying the second, extra gene. That stutter gave the mutant fish several copies of the nine-nucleotide region. The altered protein serendipitously protected the fish a bit from ice crystals, and so its progeny became more numerous in the frigid ocean.

In one of the fish descendants of the original lucky mutant, presumably, the copying machinery stuttered again, adding even more nine-nucleotide repeats and further improving the antifreeze protein. (Tandemly repeated sequences in DNA are particularly prone to being copied extra times.) The progeny of that second mutant were even more fit—they could survive in water that was marginally colder—so they quickly dominated the population. Then a deletion mutation removed the original coding region, perhaps making the antifreeze protein more stable. One or two more mutations, each of which improved it, and we've reached the modern version of the protein.

Even though we haven't directly observed it, the scenario seems pretty convincing as an example of Darwinian evolution by natural selection. It's convincing because each of the steps is tiny—no bigger than the step that yielded the sickle mutation in humans—and each step is an improvement. The original duplication that started the process happens pretty frequently. The next mutation—the stuttering that led to extra copies of the nine-nucleotide sequence—also is a type that happens relatively often (remember, stuttering is the kind of mutation that leads to Huntington's disease in humans). The next step, the deletion of the original sequence, is also not uncommon.

The likelihood of the scenario is bolstered by two other discoveries by the Illinois scientists. An unrelated group of fish from the Arctic Ocean—halfway around the world—have a gene that makes a very similar antifreeze protein, with the same repeating three amino acid residues, but which has different control regions to the left and right of the gene.[24] This suggests (but of course doesn't prove) that antifreeze proteins with the same simple repetitive sequence aren't improbable. Even more striking is that the workers found a hybrid gene from Antarctic fish that contains both the antifreeze sequence and the digestive-enzyme sequence, which they earlier had postulated was deleted in the first gene they found.[25] With the hybrid gene it really seems they had caught evolution in the act. The very kind of evolution Darwin anticipated.

SO FAR, BUT NO FURTHER

As we'll see in the next chapter, complex interactive machinery— whether in our everyday world or in a cell—can't be put together gradually. But some simple structures can. One example from our large world is a primitive dam. Because gunk accumulates, the drain in my family's kitchen sink slows and stops every so often. It doesn't much matter what makes up the garbage—bits of food, paper, big pieces and small. The gutters on our home are like that, too—pieces of different size leaves, twigs, seeds, and so on regularly plug them up. Even large rivers can get clogged by the gradual accumulation of debris. Depending on your circumstances, that might be a favorable development. Sometimes a clogged river or stream might accidentally do some animals some good if, say, it forms a reservoir. Slowing and eventually damming the flow of water doesn't require sophisticated structures—just a lot of debris. Genetic debris can accumulate in the cell, too. If it accidentally does some good, then it can be favored by natural selection. In a sense, that's what happened in the case of the Antarctic fish.

Rare examples such as the Antarctic fish set Darwinian pulses racing. But to more skeptical observers, they underscore the limits of random mutation rather than its potential. It turns out that the antifreeze protein in Antarctic fish is not really a discrete structure comparable to, say, hemoglobin. Hemoglobin and almost all other proteins are coded by single genes that produce proteins of definite length. They resemble precisely engineered dams. But the antifreeze protein is coded by multiple genes of different lengths, all of which produce amino acid chains that get chopped into smaller fragments of differing lengths—very much like the junk in my gutter. In fact, the Antarctic protein appears not to have any definitive structure. Its amino acid chain is floppy and unfolded, unlike the very precisely folded shapes of most proteins (such as hemoglobin). Nor does the antifreeze protein interact with other proteins like those found in real molecular machines discussed in the next chapter.

Like a dam across a stream, which can be made more and more effective by adding one stick or leaf or stone at a time, the job of the antifreeze protein is a very simple one, and it is relatively easy to improve the protein incrementally. It doesn't much matter whether the sticks in a dam are larger or smaller, of many different types or intermixed; as long as there are enough of them, they can block the river. And just as there are many ways to dam a river, there are many ways to make antifreeze proteins. As one group of researchers points out, "A number of dissimilar proteins have adapted to the task of binding ice. *This is atypical of protein evolution* [my emphasis]." [26]

The antifreeze protein discovered in Antarctic fish is not so much a molecular machine as it is a blood additive. Another analogy might be to a machine and the lubricant that allows it to keep running. The antifreeze protein is akin to the lubricant, which, although it might be needed for the machinery of cells to work, does not have anywhere near the complexity the machines have. In fact, to survive in the cold, plants and animals frequently add simple chemicals to their fluids similar to automobile antifreeze.[27] That works well, too.

Despite ten million years of evolution with quadrillions of fish under relentless, life-and-death selective pressure, the Antarctic antifreeze protein does not have anything like the sophistication and complexity even of such a simple protein as hemoglobin, let alone that of the stupendous, multiprotein systems that are plentiful in nature. Instead of pointing to greater things, as Darwinists hoped, the antifreeze protein likely marks the far border of what we can expect of random mutation in vertebrates.

To put matters in perspective, consider a related problem that has stumped malaria. Although malaria is a ferocious parasite, quite willing to eat anything that gets in its path, *P. falciparum* needs a warm climate to reproduce. If the temperature falls below about 65°F, the parasite slows down. When the temperature gets to 61°F, it can't reproduce. It's stymied.[28] If a mutant parasite appeared that was tolerant to somewhat lower temperatures—not to freezing conditions, just to cool temperatures—it would be able to invade regions that are now closed to it. Despite the huge number of *P. fal-*

ciparum available to mutate over thousands of years, that hasn't happened. Not all seemingly simple problems can be overcome easily, or perhaps at all.

KUDOS

Charles Darwin deserves a lot of credit. Although it had been proposed before him, he championed the idea of common descent and gathered a lot of evidence to support it. Despite some puzzles, much evidence from sequencing projects and other work points *very* strongly to common ancestry. Darwin also proposed the concept of random variation/natural selection. Selection does explain a number of important details of life—including the development of sickle hemoglobin, drug and insecticide resistance, and cold tolerance in fish—where progress can come in tiny steps.

But, although Darwin hoped otherwise, random variation doesn't explain the most basic features of biology. It doesn't explain the elegant, sophisticated molecular machinery that undergirds life. To account for that—and to account for the root and thick branches of the tree of common descent—multiple coherent genetic mutations are needed. Now that we know what sorts of mutations can happen to DNA, and what random changes can produce, we can begin to do the math to find the edge of evolution with some precision.

What we'll discover is something quite basic, yet heresy to Darwinists: Most mutations that built the great structures of life must have been nonrandom.

5

WHAT DARWINISM CAN'T DO

Debris clogging a stream shares a few things in common with the Hoover Dam. Both slow the flow of water and create large pools. Yet few people would have trouble distinguishing the two, or realizing that only one is the result of the random accumulation of twigs and leaves and mud over time. In the last chapter we looked at mutational twigs that can accumulate into a clog of biological debris. In this chapter we'll consider molecular Hoover dams.

But first, a word about complexity. In my previous book, *Darwin's Black Box,* I described certain intricate biochemical structures as "irreducibly complex" and argued that step-by-step Darwinian processes could not explain them, because they depended upon multiple parts. Critics claimed that I was simply throwing up my hands at a difficult problem, and that it would eventually be solved. They may say it again, regarding this chapter. But the discoveries of the past decade have made the problem worse, not better, both at the level of protein machinery and at the level of DNA instructions. This chapter illustrates some of the new challenges, and in the following chapters I will explain how we can generalize from them.

BOTTOM UP, TOP DOWN

I was in grammar school when the observation tower of Iacocca Hall at Lehigh University was being constructed. Busy studying the three Rs in Harrisburg about a hundred miles away, I never got the chance to see the cranes, cement mixers, dump trucks, and steel I-beams, to see all the machinery and supplies being carried around to the right places, to be joined in the right way with the right complementary pieces, to make the building where I now work. But like most of us, I've seen other buildings being constructed, so I can infer how the Iacocca Hall tower was put together. Like all such buildings, it was built in what could be called a "bottom up–top down" fashion. By bottom up I mean that of course the foundation of the building had to be poured first, the ground floor next, and so on, all the way to the zenith at the sixth floor. Successive floors have to be built on preceding ones.

By top down I mean that the building was planned. Blueprints were followed, supplies ordered, ground purchased, equipment moved in, and so on—all with the final structure of the observation tower in mind. Of course, minor features of the building might not be explicitly intended. For example, the exact color of the concrete might not matter, as long as it was an invigorating shade of gray. Or the exact placement of the handrails leading up the interior steps might not be important, as long as they were within a certain distance from the floor. Nonetheless, major structural aspects of the building were conceptualized in advance of the start of construction, and then preparations were taken to carry out the project. The need for bottom up–top down construction extends far beyond the buildings of Bethlehem, Pennsylvania. All major construction projects are conducted that way. So whenever we see a well-framed structure we may be sure it was planned.

In just the past decade or so science has unexpectedly discovered bottom up–top down construction in a location that wasn't visible just a few years earlier. It wasn't visible because the optical equipment needed to see it wasn't available. It was spotted by a powerful new microscope scrutinizing the green alga *Chlamydomonas*, a favorite laboratory organism affectionately known as Chlammy.

Since then the same type of construction has been spied in a very wide variety of cells.

In *Darwin's Black Box,* I discussed large cellular structures called the cilium and the flagellum, both of which help cells move around in liquid, acting like propellers. I had no idea how complex they really were. Both the cilium and the flagellum are big pieces of cellular machinery—big, that is, compared to the cell itself. Although they are both quite thin, their lengths can be many times longer than that of the cells to which they are attached. It turns out that the construction of big structures in the cell requires the same degree of planning—the same foresight, the same laying in of supplies, the

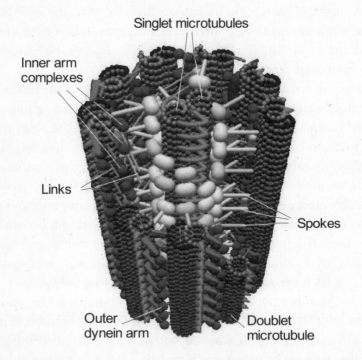

FIGURE 5.1

Computer-generated image of a section of a cilium, cut away to reveal component parts. Each small sphere is a protein of roughly the complexity of hemoglobin. The cilium is comprised of about two hundred different kinds of proteins. (Reprinted from Taylor, H. C., and Holwill, M. E. J. 1999. Axonemal dynein—a natural molecular motor. *Nanotechnology* 10:237–43. Courtesy of IOP Publishing.)

same sophisticated tools—as did the building of the observation tower at Iacocca Hall. Actually, it requires much more sophistication, because the whole process is carried out by unseeing molecular robots rather than the conscious construction workers who assemble buildings in our everyday world.

GOOSEBUMPS

In 1993 Keith Kozminski, then a graduate student at Yale, was trying out a flashy new microscope.[1] The scope had all sorts of bells and whistles, including the ability to videotape cells in real time. Kozminski focused the scope on a cilium of the single-celled alga *Chlamydomonas* and filmed what no one in the history of the world had ever seen before. Moving up one side of the cilium and down the other were a series of bumps—traveling goosebumps! A videotape of such "intraflagellar transport" (abbreviated IFT; confusingly, cilia are also sometimes called "flagella," hence "intraflagellar") can be seen on the web.[2]

Kozminski and his coworkers knew right away that there must be a lot of complex machinery behind the simple-looking, moving bumps. They hypothesized that the bumps were actually akin to traveling train cars, moving freight up the length of the cilium, and powered by various kinds of motor proteins. The bumps moved at different speeds; they went twice as fast coming back as they did going out from the cell to the tip of the cilium. So the investigators deduced that there were two separate mechanisms responsible for the outward trip and the return. Switching from videotapes to still pictures taken by higher-resolution microscopes, the workers were able to make out some details of the bumps. They saw groups—later called "rafts"—of up to forty lollipop-shaped particles situated between the outer circumference of the protein part of the cilium and the membrane that encloses it. Unlike some other types of transport machinery in the cell, the lollipops were not "vesicles." That is, they were not enclosed spaces wrapped by a protein or membranous coat.

With a combination of serendipity and skill, a window was opened onto elegant and unsuspected cellular machinery. In the decade since Keith Kozminski first glimpsed IFT, tremendous progress has been made in detailing the many protein players in the mechanism, as well as the often dire consequences when mutations disrupt its work. Over the next several pages we'll look at some details of IFT.

BUILDING A TOWER

IFT is the machinery that builds and maintains the cilium. If a cilium is cut off a Chlammy cell, another one will be generated over the course of an hour or so. During that hour little IFT rafts can be spotted busily flowing up one side of the growing structure and down the other. If, however, by clever laboratory manipulations, one or more of the protein components of IFT are deliberately broken, an amputated cilium will no longer be rebuilt.

In bottom up–top down construction, for convenience materials are often gathered in advance and brought to the building site. That was surely the case for Iacocca Hall's tower, and it's also the case for the cilium. Before starting to build a new cilium, cellular materials are brought to a staging site near the bottom of what will be the new structure. Of course, in human construction projects the conscious workers know which materials they need, recognize them, and bring only needed materials into the building site. In the cell, however, that all has to be done by highly sophisticated, automated mechanisms. It had been hypothesized that, in an area near the base of the new cilium, things called "transition fibers" act as filters to keep out unwanted, potentially disruptive materials. Douglas Cole of the University of Idaho reasoned that if that were indeed the case—if construction materials needed an admission ticket to get into the cilium—then no new materials would be allowed past the transition zone into the cilium if IFT were experimentally halted. That is precisely what was seen in several Chlammy mutants.[3] The exact details of the filtering mechanism aren't yet known, but you can be sure they won't be simple.

Like all analogies, the comparison of the building of a cilium to a human construction project fails in a number of respects, all of which emphasize the much greater sophistication of cellular construction. Here I'll mention just one aspect. Although a human construction crew leaves a building project once it's completed, that's not the case with the cell. If IFT is experimentally interrupted in a cell that already has a full, finished cilium, the cilium immediately starts to shorten until it disappears. IFT continues throughout the lifetime of the cilium, not only constantly bringing in new copies of ciliary components, but also removing old material. Experiments have shown that in apparently stable cilia whose length remains constant, in a period of several hours over eighty different kinds of proteins amounting to 20 percent of the mass of the cilium are exchanged.[4]

The current model for IFT pictures the freight cars at the beginning of construction to be mostly full. (Figure 5.2) After construction is completed the trains keep coming at about the same rate, but now some of the cars are empty. Apparently some as-yet-unknown switching mechanism senses how much material the cilium needs

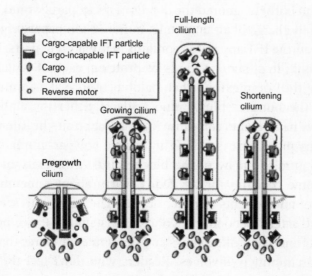

FIGURE 5.2

Intraflagellar transport (IFT). Molecular containers carry protein cargo from the cell to the tip of the flagellum. Containers return empty. To maintain the correct length of the cilium after it is built, a greater percentage of containers are believed to switch from a "cargo-capable" to a "cargo-incapable" form. (Reprinted from Snell, W J., Pan, J., and Wang, Q. 2004. Cilia and flagella revealed: from flagellar assembly in Chlamydomonas to human obesity disorders. *Cell* 117:693–97. Courtesy of Elsevier Publishing.)

at any particular moment and changes the proportion of freight cars between "cargo-capable" and "cargo-incapable" as the need arises. Unlike the tower of Iacocca Hall, the cilium is a dynamic structure, in which many of its protein parts are actively altered in response to changing internal and external conditions.

THE FULL MONTY

Writing of IFT as using little "train cars" shaped like "lollipops" that run along molecular "railroad tracks" is of course baby talk. The baby talk has a serious purpose—to abstract some important, over-arching points without getting bogged down for the moment in too many details. But the moment comes when details have to be fully faced. A real train, say a steam locomotive, contains very many parts that all have to be working in order for the train to operate. An engineer who blithely ignored the details of those parts would soon find himself in charge of an immobile, hundred-ton paperweight. In the same way, the IFT apparatus contains many protein parts. It directly contains at least sixteen kinds of proteins, each of which is itself roughly the complexity of hemoglobin. And just as a mutation in one of the hundred-plus amino acids of either the alpha or beta chains of hemoglobin can cause it to malfunction, the same is true of the many protein parts of IFT. In the next few paragraphs we'll stare directly into the maw of the biochemical complexity of IFT, and then come back up for air. Don't worry about remembering the names of components or other details. The point is to see how elegant and interdependent the coherent system is—to see how different it is from the broken genes and desperate measures that random mutation routinely involves. Readers who don't feel the need for this level of detail may wish to skip to the next section.

Biochemical studies show that IFT can be conceptually broken down into several parts. The first part consists of the motor proteins that carry the IFT particles along the interior of the cilium. The motor protein that carries the particle toward the tip of the cilium is different from the one that carries it back. The trip out is powered by kinesin-II, one member of a family of kinesin motor proteins that perform a variety of jobs in the cell. Kinesins come in a range of structural variants. Kinesin-II is found only in cells that have IFT, but not in cells such as those of yeasts and higher plants that don't. (Yeasts and higher plants don't have cilia.) One study showed that cells that contain a mutant, fragile kinesin-II can form cilia at lower temperature (about 68°F) where the mutant protein works. But at higher temperature (90°F) where the protein is unstable, IFT stopped and cilia began to be resorbed. The trip back is powered by a dynein motor protein. When a mutant, disabled dynein was placed in Chlammy cells that didn't have cilia, new cilia that were formed were very short and bulging with IFT particles that contained kinesin-II. Apparently, the machinery for getting particles in was working fine, but the machinery for getting particles back out was broken, so the incipient cilium became overstuffed. Exactly what causes IFT to shift from kinesin-powered transport to dynein transport at the tip of the cilium remains unknown.

The second conceptual part of IFT is called the IFT particle. It's the container that grabs hold of the correct proteins to be carried in or out and releases them at the proper point. The IFT particle consists of sixteen separate proteins that bind together in one aggregate. Under some experimental conditions the sixteen-protein complex can be separated into two complexes—called A and B—that contain six and ten proteins respectively. It's not certain, but it seems that complexes A and B may play distinct roles in the cell.[5] The proteins of complexes A and B contain substructures that are known to be particularly good at binding diverse proteins—exactly what you need to transport the many kinds of protein cargo that travel by IFT along the cilium.

TRAIN WRECK

When parts of a railroad transportation system are missing or broken—when a railroad tie is misaligned, a rivet or two missing, a bolt holding a wheel on the engine broken—disaster may not be far behind. So, too, with IFT. Cilia aren't just oars flapping in the water—they participate in a wide range of critical biological functions. If they aren't well maintained, a lot of things can go wrong. In the past decade defects in IFT have been shown to affect a number of important processes.

The earliest hint that cilia have a number of hidden but vital tasks came in the mid-1970s when Swedish scientist Björn Afzelius reported the cases of four men who suffered from infertility and chronic sinusitis. Since the tail of a human sperm is a modified cilium that powers its swimming, and since ciliated cells line the sinus cavities, Afzelius examined respiratory tissue from the patients and checked their cilia. Although cilia were there, they lacked the dynein that's present in normal cilia, and thus were unable to move. Afzelius also noted something odd about his patients—several of them had *situs inversus*, that is, their hearts were on the right sides of their bodies and their livers on the left, the opposite of normal. Afzelius's observation suggested that anything that broke a cilium might cause the left-right mixup. In 1999 some Japanese workers genetically manipulated mice to be missing one of the proteins that forms the kinesin motor of IFT. The mice died before birth. Examination of the embryos showed many to have *situs inversus*. So one conclusion is that a properly working IFT is necessary for correct embryonic development.[6]

Another area affected by IFT is vision. In the photoreceptor cells of the retina of vertebrates, a large inner segment (IS) is connected by a thin neck to a large outer segment (OS). The IS harbors the guts of the cell—nucleus, ribosomes, and so on—while the OS has the specialized machinery dedicated to vision. Since all the supplies needed for the OS are first constructed in the IS, they have to be shuttled from one compartment to the next. The connecting neck is actually a modified, nonmotile cilium, so it is suspected that supplies reach the OS by IFT. That hypothesis has been strengthened by recent work showing that a mutation in just one of the sixteen IFT proteins in mice causes the rodent retinas to degenerate.[7] In another study the kinesin IFT motor was intentionally broken in lab mice; proteins that normally are shipped out to the OS became stuck in the IS. Eventually, as many improperly functioning cells do, the defective photoreceptor cells activated their self-destruct program and committed suicide.[8]

People who suffer from polycystic kidney disease develop large cysts on their kidneys (and other organs, too) and gradually lose kidney tissue, leading to kidney failure. Since kidneys are necessary to filter blood, the consequences can be deadly. Studies on humans showed that mutations in the genes for either of two proteins, called polycystin-1 and polycystin-2, were associated with the disease. Polycystin-2 is found in certain cilia of kidneys. In experiments with mice, deliberate breaking of one of the proteins of IFT eliminated the construction of those cilia and led to polycystic kidney disease in the animal.[9] The conclusion is that IFT is needed for proper kidney function, too.

Besides its role in embryo development and eye and kidney function, IFT likely plays a number of other roles in the cell. Experiments point to functions in sensing the concentration of chemicals in liquid (osmotic sensing), receiving chemical signals, mating behavior in worms, and more.

IRREDUCIBLE COMPLEXITY SQUARED

When we're children, life seems simple. We don't know how the world really works, and don't even know enough to ask questions about it. Our parents meet all our needs; our country can do no wrong; our school is the best. But while growing up, most of us discover that things aren't so straightforward as they first appeared. A school bully punches us in the nose; we hear some of our country's actions denounced by people whose opinions we respect; Dad tells us we have to earn the money ourselves—he won't just give us a car. Life gets more and more complicated. So, too, with biochemistry. In Darwin's era in the nineteenth century the cell seemed boringly simple. The eminent embryologist Ernst Haeckel called it a "simple little lump of albuminous combination of carbon"[10]—in other words, just some gray goo. As it grew up over the years science has learned that the cell is tremendously more complex than Haeckel thought.

Now we realize that the cilium, too, is tremendously complex. Now we know that a cilium is more than just a flapping oar, useful for swimming or keeping liquid moving through a tissue. It's also a sophisticated chemical sensor involved in a wide array of biological processes. It is dynamic in multiple, independent ways—not just mechanically dynamic, but also functionally dynamic, continuously being rebuilt to better reflect and respond to its environment.

And now the problem of its irreducible complexity has been enormously compounded. Let's reconsider the mousetrap—the paradigm of irreducible complexity I discussed in *Darwin's Black Box*. A standard mechanical mousetrap needs multiple parts to work. If the spring is removed or a metal bar broken, the trap won't catch any mice. Despite the imaginative but dubious efforts of Darwin fans over the past decade,[11] it's extremely difficult to see how something like a mousetrap could actually evolve by something akin to a blind Darwinian search process. But now let's move beyond the structure of just the mousetrap itself. Imagine an automated mousetrap factory that assembled the parts of the trap, set it, and reset the trap each time it went off. Clearly, the complexity of such a system is much greater than the complexity of the mousetrap alone. And just as the odds against winning a Powerball lottery skyrocket the more numbers you have to match, the difficulty of explaining how a mousetrap-making system could arise by "numerous, successive, slight modifications" (as Darwin required of his theory) rises exponentially the more separate kinds of parts the system contains.

IFT exponentially increases the difficulty of explaining the irreducibly complex cilium. It is clear from careful experimental work with all ciliated cells that have been examined, from alga to mice, that a functioning cilium requires a working IFT.[12] The problem of the origin of the cilium is now intimately connected to the problem of the origin of IFT. Before its discovery we could be forgiven for overlooking the problem of how a cilium was built. Biologists could vaguely wave off the problem, knowing that some proteins fold by themselves and associate in the cell without help. Just as a century ago Haeckel thought it would be easy for life to originate, a few decades ago one could have been excused for thinking it was probably easy to put a cilium together; the pieces could probably just glom together on their own. But now that the elegant complexity of IFT has been uncovered, we can ignore the question no longer.

How do Darwinists explain the cilium/IFT? In 1996 in *Darwin's*

Black Box I surveyed the scientific journals and showed that very few attempts had been made to explain how a cilium might have evolved in a Darwinian fashion—there were only a few attempts. Although Brown University biologist Kenneth Miller argued in response that the two-hundred component cilium is not really irreducibly complex, he offered no Darwinian explanation for the step-by-step origin of the cilium. Miller's professional field, however, is the study of the structure and function of biological membranes, and his rejoinder appeared in a trade book, not in the scientific literature. An updated search of the science journals, where experts in the field publish their work, again shows no serious progress on a Darwinian explanation for the ultracomplex cilium.[13] Despite the amazing advance of molecular biology as a whole, despite the sequencing of hundreds of entire genomes and other leaps in knowledge, despite the provocation of *Darwin's Black Box* itself, in the more than ten years since I pointed it out the situation concerning missing Darwinian explanations for the evolution of the cilium is utterly unchanged.[14]

On the origin of the cilium/IFT by random mutation, Darwinian theory has little that is serious to say. It is reasonable to conclude, then, that Darwinian theory is a poor framework for understanding the origin of the cilium.

The cilium is no fluke. The cell is full of structures whose complexity is substantially greater than we knew just ten years ago. (In Appendix C, I discuss intricacies of the bacterial flagellum and its construction, for readers who enjoy plenty of details.) The critical question is, of course, Can mutation of DNA explain this? Or rather, can *random* mutation explain it? Life descended from a common ancestor, so DNA did mutate—change from species to species. But what drove the crucial changes?

Repeating Darwin's own mistakes, modern Darwinists point to evidence of common descent and erroneously assume it to be evidence of the power of random mutation.[15] Yet if modern malaria can't deal with the single amino acid change of sickle hemoglobin, why should we think that the IFT system would be supplied by random mutation in some ancient cell? If the human genome is substantially harmed by its trench warfare with *P. falciparum*, why do we think competition would build an elegant molecular outboard motor? To ask such questions is to answer them. There is no evidence that Darwinian processes can make anything of the elegance and complexity of cilia.

TIMING IS EVERYTHING

If the cilium is likened to the tower of Iacocca Hall, then IFT can be compared to the bulldozers, cranes, and other machinery needed to construct it. But that's not all that's needed for bottom up–top down construction. To appreciate the massive challenge that cellular systems present to random mutation, we have to consider more than just physical features, more than the final structures themselves and the construction machinery needed to build them. We also have to consider the molecular *planning* that goes into the project. Genetic control of planning is in some ways the most difficult aspect of a molecular construction process for scientists to investigate, but is no less critical than the physical parts that make up the final structure.

A large construction project has to be conducted in an orderly manner. Orderly construction isn't needed because of some aesthetic obsession with neatness; it's needed because if there are too many machines and other items on the construction site they can interfere with each other. If all the items needed for a finished office building were present on site from the start, they would get in each other's way; some would be damaged, machinery might be clogged. If office furniture were scattered over the construction site at the start, at the same time when steam shovels first arrived to dig the foundation, the furniture would likely get scooped up in a shovel or crushed under a tractor tread. The end result would be a mess.

Physical construction in the cell is almost exclusively the job of proteins. Proteins constitute the molecular bulldozers, steam shovels, train engines, train cars, railroad tracks, and all the other tools, both large and small, needed for construction projects. Of course, the genes that code for the proteins are composed of DNA, so ultimately *all* the information needed to make *all* the material required for construction—both the construction machinery and the materials that make up the office tower itself—resides in DNA. In addition to those genes, however, the DNA of a cell also has regions that act as control signals. The control signals of DNA, in conjunction with control proteins, orchestrate the project, to make sure that the proper machinery is made at the proper time in the proper amounts.

Elucidating how the cell functions is very difficult work, and much remains unknown. Although aspects of IFT have been unveiled in the past decade, the control program for making a cilium is still largely a mystery. However, in that same time remarkable progress has been made in outlining the control program for another large structure, the bacterial flagellum (see Appendix C). Briefly, the bacterial flagellum is an outboard motor that bacteria use to swim. In order to illustrate the planning that molecular construction must involve, over the next few paragraphs I'll describe what has recently been learned about the control of flagellum construction. (Some readers may wish to skip to the next section.)

Just as the outboard motor of a motorboat in our everyday world consists of a large number of parts (propeller, spark plugs, and so on), so does the molecular outboard motor. The flagellum has dozens of protein parts that do the particular jobs necessary for the complex system to work. Those dozens of proteins are coded by dozens of genes in a bacterial cell. The genes are grouped into fourteen bunches called "operons." Next to each operon in the DNA are control signals. The control signals themselves fall into three categories we'll call class 1, class 2, and class 3. The genes for proteins that have to be made first in the construction process have class 1 control signals, those genes that go second have class 2 signals, and so on.

Most of the time, a bacterial cell isn't building a flagellum, because it already has one. However, after cell division a new cell has to start the construction program. To begin, the DNA control regions for class 1 genes mechanically "sense" that the time has come and switch on class 1 genes. There is just one operon in class 1, which contains just two genes. The genes code for two protein chains, which, like the alpha and beta chains of hemoglobin, stick to each other to make a single functioning protein complex. That protein is neither a part of the flagellum nor a part of the construction machinery. Rather, it's akin to the foreman of a project, who has to tell the other workers what to do. Let's call it the "boss" protein.

The boss protein binds specifically to the DNA control regions of the seven class 2 operons, mechanically turning them on. Class 2 genes code for the proteins that make up the foundation of the flagellum (plus some helper proteins), just as you'd expect in bottom-up construction. One class 2 gene, however, isn't part of the foundation. It's another control protein. Let's call it the "subboss" protein. The subboss protein binds to the DNA control region of class 3 genes, which comprise proteins that make the outer parts of the flagellum. So each class of genes contains the gene for a protein that will turn on the next class.

But that's not all. Clever as that part is, the control system is much more finely tuned than just the cascading control proteins. For years researchers knew that if the genes for any of a score of protein parts in class 2—the ones that made up the foundation of the flagellum—were experimentally broken in the lab, the genes for the outer parts of the flagellum would remain switched off. But how could so many genes all control later construction?

Class 3 contains a gene for a protein that binds tightly to the subboss protein, inactivating it. Let's call that the "checkpoint" protein. Why turn on the subboss only to immediately inactivate it with the checkpoint protein? Later in the construction project, a clever maneuver gets rid of the checkpoint protein. The flagellum not only is an elegant outboard motor, but also contains a complex pump in its foundation, which actively extrudes class 3 protein parts to form the outer portion of the structure.

Here's the elegant trick. When the pump in the foundation of the flagellum is completed and running, one of the first proteins to be extruded is the checkpoint protein. Getting rid of the checkpoint protein releases the subboss protein to bind to the control regions of class 3 operons, switching on the genes for the outer portion of the flagellum. So the completion of the first part of the flagellum is directly linked to the switching on of the genes to make the final parts of the flagellum.

"MIND-BOGGLING COMPLEXITY"

In just the past few years a group of Israeli scientists has developed clever new laboratory techniques to analyze in even finer detail the control exerted by DNA control elements on the construction of the flagellum. By successively joining the control elements to the gene for a protein that can be detected by its fluorescence, the scientists showed that, even within classes 2 and 3, the control elements switch the genes on in the order that they are needed for construction. Within class 2, the genes needed for the bottom of the foundation are switched on before the genes for the top of the foundation, and within class 3, genes for the bottom of the top are activated before genes for the top of the top.[16]

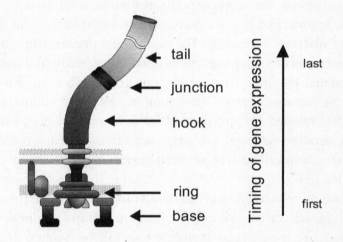

FIGURE 5.3

Genes for the construction of the bacterial flagellum are activated in a precisely timed fashion. Those needed for construction of the bottom of the molecular machine are switched on first, followed in order by those needed for more distant parts. (Illustration of the flagellum reprinted courtesy of the Kyoto Encyclopedia of Genes and Genomes, Kanehisa, M., Goto, S., Hattori, M., Aoki-Kinoshita, K. F., Itoh, M., Kawashima, S., Katayama, T., Araki, M., and Hirakawa, M. 2006. From genomics to chemical genomics: new developments in KEGG. *Nucleic Acids Res.* 34:D354–57.)

The same group of scientists has examined DNA control elements for other cellular systems and discovered similar elegance there. When they studied cellular biochemical pathways for making amino acids, they discovered what is called "just-in-time" organization, where a protein is made as close to the time it's needed as possible:

> Mathematical analysis suggests that this "just-in-time" transcription program is optimal under constraints of rapidly reaching a production goal with minimal total enzyme production. Our findings suggest that metabolic regulation networks are designed to generate precision promoter timing and activity programs that can be understood using the engineering principles of production pipelines.[17]

What does all this jargon mean? Simply put, the more closely we examine the cell, the more elegant and sophisticated we discover it to be. Complex, functional structures such as the cilium and flagellum are just the beginning. They demand intricate construction machinery and control programs to build them. Without those support systems, the final structures wouldn't be possible. The bacterial flagellum contains several dozen protein parts. The cilium, which so far has resisted investigation of its DNA control program, has several hundred. There is every reason to think that the control of its construction will have to be much more intricate than that of the flagellum.

Control of construction projects and other activities in the cell is difficult for scientists to investigate, because "control" is not a physical object like a particular molecule that can be isolated in a test tube. It's a matter of timing and arrangement. The upshot is that even now in the twenty-first century—more than fifty years after the double helical shape of DNA was discovered by Watson and Crick, and decades after the first X-ray crystal structures of proteins were elucidated—science is still discovering fundamental new mechanisms by which the operation of the cell is controlled.

Recently—some sixty-five years after George Beadle and Edward Tatum proposed the classic definition of a gene as a region of DNA that codes for an enzyme—an issue of the journal *Nature* ran a feature with the remarkable title "What Is a Gene?" The gist of the article was that the control systems that affect when, where, and how much of a particular protein is made are becoming so complex, and their distribution in the DNA so widespread, that the very concept of a "gene" as a discrete region of DNA is no longer adequate. Marvels the writer, "The picture these studies paint is one of mind-boggling complexity." [18]

DELIMITING THE EDGE

Where is it reasonable to draw the edge of evolution? In this chapter and the preceding one I intended to circumscribe that question— show examples of what I think clearly can and what clearly cannot be explained by random mutation and natural selection. Some- where between those extremes, then, lies the edge.

On the one side are our very best examples—from humanity's trench war with parasites—of what random mutation and natural selection are known to do. We know that single changes to single genes can sometimes elicit a significant beneficial effect. The classic example, taught in virtually all biology textbooks, is that of sickle cell hemoglobin, where a change of one amino acid confers resist- ance to malaria, saving many children from premature deaths. Other examples fit the single-change profile, such as HbC and HbE, warfarin and DDT resistance, and so on. Random mutation also produced a long list of broken genes that can be beneficial in dire circumstances: thalassemia, G6PD deficiency, CCR5 deletion, and so on.

More rarely, several mutations can sequentially add to each other to improve an organism's chances of survival. An example is the breaking of the regulatory controls of fetal hemoglobin to help alle- viate sickle cell disease. Very, very rarely, several amino acid muta- tions appear simultaneously to confer a beneficial effect, such as in chloroquine resistance with mutant PfCRT. Changing multiple amino acids of a protein at the same time requires a population size of an enormous number of organisms. In the case of the malarial parasite, those numbers are available. In the case of larger creatures, they aren't.

On the other side are the examples of what random mutation and natural selection clearly cannot do. In this chapter I discussed several illustrations—IFT and the control of bacterial flagellum construction—of the kind of astonishingly complex, coherent systems that fill the cell. Those systems aren't built from just one or two amino acid changes to random proteins of systems doing other jobs—they consist of dozens of different proteins dedicated to their tasks. They didn't arise by breaking genes; they required the coordinated construction of many new genes. Cilia and flagella are not only stupendously complex systems in their own right, but they have complicated systems dedicated to their construction, and genetic control systems coordinating that construction, whose intricacy science is only now beginning to appreciate.

The structural elegance of systems such as the cilium, the functional sophistication of the pathways that construct them, and the total lack of serious Darwinian explanations all point insistently to the same conclusion: They are far past the edge of evolution. Such coherent, complex, cellular systems did not arise by random mutation and natural selection, any more than the Hoover Dam was built by the random accumulation of twigs, leaves, and mud.

6

BENCHMARKS

It's time to consider some general principles. How do we decide if some biological feature is unlikely to have been produced by random mutation and natural selection? Writing of other matters in their book *Speciation,* evolutionary biologists Jerry Coyne and Allen Orr pinpoint the key principle:

> The goal of theory, however, is to determine not just whether a phenomenon is theoretically possible, but whether it is *biologically reasonable*—that is, whether it occurs with significant frequency under conditions that are likely to occur in nature.[1]

In this book we'll apply the paramount Coyne-Orr principle to Darwinian evolution as a whole (which they do not).[2] In light of the recent tremendous progress of science, can we determine not what is merely theoretically possible for Darwinian evolution, not what may happen only in some fanciful Just-So story, but rather what is *biologically reasonable* to expect of random mutation and natural selection at the molecular level? If we can decide what is biologically reasonable to expect of unguided evolution, then we can also determine what is unreasonable to expect of it.

Since we'll be looking at borderline, marginal cases, determining the ragged edge of evolution will necessarily be more tentative than finding clear-cut examples of what certainly can and cannot be done by Darwinian processes. Sickle hemoglobin can inarguably be explained by mutation and selection, the bacterial flagellum cannot. Is the edge of Darwinian evolution closer to sickle hemoglobin, or closer to the flagellum?

In this chapter I develop two criteria by which to judge whether random mutation hitched to natural selection is a biologically reasonable explanation for any given molecular phenomenon. The criteria, spelled out in more detail over the rest of the chapter, are the following.

- First, *steps*. The more intermediate evolutionary steps that must be climbed to achieve some biological goal without reaping a net benefit, the more unlikely a Darwinian explanation.
- Second, *coherence*. A telltale signature of planning is the coherent ordering of steps toward a goal. Random mutation, on the other hand, is incoherent; that is, any given evolutionary step taken by a population of organisms is unlikely to be connected to its predecessor.

I discuss evolutionary steps over the next three sections, and coherence in the subsequent three. In this chapter, we'll continue to examine the molecular level of life. Later, we'll extend the analysis to higher levels.

MONKEYS, TYPEWRITERS

A few years ago a curious fellow decided to test the old saw that, given typewriters and enough time, an army of monkeys would eventually produce the works of Shakespeare. A computer with keyboard was placed in a cage containing six macaques in a British zoo and left for four weeks. The result? "The macaques—Elmo, Gum, Heather, Holly, Mistletoe and Rowan—produced just five pages of text between them, primarily filled with the letter S. There were greater signs of creativity towards the end, with the letters A, J, L and M making fleeting appearances, but they wrote nothing even close to a word of human language."[3] The five pages have been published under the ironic title "Notes towards the Complete Works of Shakespeare."

Because the names of amino acids that constitute the building blocks of proteins are abbreviated as single letters (L for leucine, S for serine, and so forth), as are the nucleotides that are the building blocks of DNA (A, C, G, and T) proteins can be likened to words and paragraphs, and the information contained in human DNA can be likened to an encyclopedia. Extending the analogy, evolution can be pictured as monkeys at a typewriter—not actually writing the words from scratch, but occasionally introducing a spelling change at random into a pre-existing text. If the change is a misspelling or ungrammatical, it is tossed out. However, if the spelling change leads to some new meaning, then (the analogy suggests) natural selection might preserve it. This analogy is crude and strained, but it has been popular among Darwinian popularizers. For example, in *Darwin's Dangerous Idea,* the philosopher Daniel Dennett considers the first line of the classic novel *Moby Dick:* "Call me Ishmael." A change or insertion of just one character can change the sense of the sentence, notes Dennett. For example, inserting a comma gives "Call me, Ishmael," making it seem as if another person is addressing Ishmael, rather than he himself speaking. Switching another letter, writes Dennett, yields "Ball me Ishmael," changing the meaning drastically.

For now let's overlook the fact that neither of these changes, which redefine Ishmael as someone other than the narrator, fits easily with the point of view of the next sentence ("Some years ago—never mind how long precisely—having little or no money in my purse, and nothing particular to interest me on shore, I thought I would sail about a little and see the watery part of the world") or the rest of the book. Let's just concentrate on that first sentence. The critical point for our present purpose is that a change of one character—the addition of a comma, the switch of a C for a B—can alter the meaning of the sentence. As with written sentences, so too with biology. The change of a single amino acid or nucleotide "character" in protein or DNA can alter its "meaning"—its biological activity—as it does for sickle hemoglobin.

The eminent evolutionary biologist John Maynard Smith, who

died in 2004, addressed this point over thirty years ago and reached an important conclusion.

> The model of protein evolution I want to discuss is best understood by analogy with a popular word game. The object of the game is to pass from one word to another of the same length by changing one letter at a time, with the requirement that all the intermediate words are meaningful in the same language. Thus WORD can be converted into GENE in the minimum number of steps as follows:

> WORD   WORE   GORE   GONE   GENE

Because mutations are relatively rare, the monkey's typing is almost always judged after a single keystroke. Bad changes are quickly eliminated. So, reasoned Smith, evolution has to slog along one tiny, beneficial step at a time. If it needs two changes to help, it gets stuck. University of Rochester evolutionary biologist H. Allen Orr recently seconded John Maynard Smith's reasoning:

> Given realistically low mutation rates, double mutants will be so rare that adaptation is essentially constrained to surveying—and substituting—one-mutational step neighbors. Thus if a double-mutant sequence is favorable but all single amino acid mutants are deleterious, adaptation will generally not proceed.[5]

If two mutations have to occur before there is a net beneficial effect—if an intermediate state is harmful, or less fit than the starting state—then there is already a big evolutionary problem. For example, changing "Call me Ishmael" to "Call me Israel" might be beneficial in some context, but it would require several changes, and thus would appear to be beyond what Smith and Orr allow. Yet Smith and Orr actually overstate the case. Although multiple evolutionary changes are unlikely, as we saw in Chapter 5 they can occur, at least for the prolific malarial parasite. Several changes apparently did occur together very infrequently in the protein that conferred resistance to chloroquine. So the Smith-Orr criterion of two changes, while reasonable as a rule of thumb, is not a hard and fast law. But what allows exceptions to the rule of thumb? And what if instead of two mutations, three mutations were needed at once? Or more?

CLIMBING THE TOWER

As I mentioned earlier, I work in Iacocca Hall at Lehigh University's branch location known as the "Mountaintop Campus." One prominent feature of Iacocca Hall is an observation tower that stands about six stories high. At the top of the tower is a room with large windows all around. On a clear day you can see all the way east to New Jersey, west to Allentown, and north to the Poconos. It's a great room for university dinners, and it's rented out occasionally for wedding receptions. Elevators lead from the third floor to the tower room; there are no intermediate floors. If the elevators are out of service, the only access is through a narrow staircase.

Suppose you were standing outside the tower with a friend, and he asked how long it would take to reach the top. Jokingly, you might say that, from the outside, it'd take forever, because no one could walk up the sheer outside walls. Even with a running jump, the best athlete on earth couldn't get up more than a small portion of the height. Ha, ha, the friend would respond, what a droll fellow you are. But he knows there are stairs inside, and wanted to know the time needed to climb the stairs.

For me, walking up the stairs to the tower might take ten minutes. For a younger person who's in somewhat better shape, maybe one minute. In either case, it's just a moment of time compared to the struggle faced by a jumper on the outside of the building. Without stairs (or a ladder or mountain-climbing equipment and so on) the tower room is effectively beyond reach. Only breaking the climb into many small steps with a staircase makes the task possible.

Suppose, though, that a person walking up the inside stairs encountered a missing step, so to get to the next step he had to climb twice the normal step-to-step distance. How long would it then take to climb the stairs? It would depend. If the climber were a frail old man, the missing step might be equivalent to a brick wall—virtually impassable. If the climber were just an out-of-shape, middle-aged couch potato (like me), one missing step wouldn't be too much of a problem. Swinging a leg up to the next step might induce some puffing and wheezing, but could be done, albeit slowly. If the climber were an athletic twenty-something, she would bound over the break. But suppose two steps were missing, or three. With three missing steps, the couch potatoes would likely be left behind with the frail old men, but the jocks could still go on. With more missing steps, even the athletes would have trouble. Some might need multiple tries before successfully making the jump, so their progress would be slower. If a whole flight of steps were missing between floors, then even the athletic twenty-somethings would be stymied. At some point, with enough steps missing, even the most athletic person on the planet couldn't pass.

Of course the stairway to the tower room in Iacocca Hall is an analogy for Darwinian evolution, one that presents the problem for evolution in a different way than monkeys and typewriters do. As with scaling physical heights, so too with ascending biological heights. If there are many closely spaced steps leading from one level of biology to another, then moving between them is trivial. If there are no such steps, the task is effectively impossible. Charles Darwin realized the distinction from the beginning. In *The Origin of Species* he emphasized that his new theory of evolution by natural selection had to explain changes in biological systems by "numerous, successive, slight modifications" of old ones. He insisted that, for his theory to be correct, evolution had to be a gradual, step-by-tiny-step process leading from one working arrangement to a new working arrangement through a series of intermediate states, all of which also worked, and all of which were just a small biological distance from their preceding and succeeding steps. Since Darwin lived before the discovery of the molecular basis of life, he didn't realize, as John Maynard Smith and Allen Orr did, that those steps were actually tiny changes in molecules. Darwin knew that if there were steps between biological levels, his idea would work. He also knew that if the stairs were missing, "my theory would absolutely break down."

Random mutation is the perfect tool for the evolutionary job when steps are continuous and close together. When there are some broken stairs, with small gaps between steps, it's a potential tool. The seriousness of the breach in the steps depends on the health of the climber. In evolutionary terms, roughly, the larger the population of a species, the "healthier" it is. Species with tiny populations are the frail old men of biology, foiled by a missing step. Ones with abundant population are the athletes, able to leap multiple missing steps. Yet, as with human athletes and missing stairs, there comes a point where even the most abundant population on earth cannot jump an evolutionary barrier. Random mutation is almost certainly useless, even for the largest populations, when a flight of stairs is missing between biological floors.

GOING STRAIGHT

The number of missing steps between one biological structure and another is the first major criterion I'll use for drawing a line marking the edge of Darwinian evolution. Recall the example of sickle cell disease. The sickle cell mutation is both a life saver and a life destroyer. It fends off malaria, but can lead to sickle cell disease. However, as we saw in Chapter 2, hemoglobin C-Harlem has all the benefits of sickle, but none of its fatal drawbacks. So in western and central Africa, a population of humans that had normal hemoglobin would be worst off, a population that had half normal and half sickle would be better off, and a population that had half normal and half C-Harlem would be best of all. But if that's the case, why bother with sickle hemoglobin? Why shouldn't evolution just go from the worst to the best case directly? Why not just produce the C-Harlem mutation straightaway and avoid all the misery of sickle?

The problem with going straight from normal hemoglobin to hemoglobin C-Harlem is that, rather than walking smoothly up the stairs, evolution would have to jump a step. C-Harlem differs from normal hemoglobin by two amino acids. In order to go straight from regular hemoglobin to C-Harlem, the right mutations would have to show up simultaneously in positions 6 and 73 of the beta chain of hemoglobin. Why is that so hard? Switching those two amino acids at the same time would be very difficult for the same reason that developing resistance to a cocktail of drugs is difficult for malaria—the odds against getting two needed steps at once are the multiple of the odds for each step happening on its own.

What are those odds? Very low. The human genome is composed

of over three billion nucleotides. Yet only a hundred million nucleotides seem to be critical, coding for proteins or necessary control features. The mutation rate in humans (and many other species) is around this same number; that is, approximately one in a hundred million nucleotides is changed in a baby compared to its parents (in other words, a total of about thirty changes per generation in the baby's three-billion-nucleotide genome, one of which might be in coding or control regions).[6] In order to get the sickle mutation, we can't change just any nucleotide in human DNA; the change has to occur at exactly the right spot. So the probability that one of those mutations will be in the right place is one out of a hundred million. Put another way, only one out of every hundred million babies is born with a new mutation that gives it sickle hemoglobin. Over a hundred generations in a population of a million people, we would expect the mutation to occur once by chance. That's within the range of what can be done by mutation/selection.

To get hemoglobin C-Harlem, in addition to the sickle mutation we have to get the other mutation in the beta chain, the one at position 73. The odds of getting the second mutation in exactly the right spot are again about one in a hundred million. So the odds of getting both mutations right, to give hemoglobin C-Harlem in one generation in an individual whose parents have normal hemoglobin, are about a hundred million times a hundred million (10^{16}). On average, then, nature needs about that many babies in order to find just one that has the right double mutation. With a generation time of ten years and an average population size of a million people, on average it should take about a hundred billion years for that particular mutation to arise—more than the age of the universe.

(Some readers might be wondering if this is a fair analysis of double mutations in general, since I'm focusing just on the set that gives C-Harlem, yet other sets of mutations might arise in nature that might be as helpful as C-Harlem. It turns out that consideration won't affect matters much. First, as we saw with the response of malaria to chloroquine, there may be very few useful evolutionary responses possible, even taking two steps at a time. Out of a hundred billion billion parasites, only one effective response was produced, a change in PfCRT. Such limited ability to respond is also seen in resistance to warfarin by rats and the similar responses of flies and mosquitoes to insecticides, and appears to be typical. Second, and more important, the odds against obtaining a cluster of mutations increases exponentially the more sites that have to be matched, but decreases only linearly with the number of combinations that are helpful. Even if there were a hundred possible double mutations that would help, that would decrease the average waiting time in the example above only linearly, just by a factor of a hundred, from a hundred billion years to a billion years. The general point would remain, that the need to mutate two or more sites together to get an effective evolutionary response immediately makes the problem much more difficult than having to match just one.)

Of course, C-Harlem did arise, relatively recently, in New York City. It happened exactly the way Darwin envisioned, by "numerous [well, two, anyway], successive, slight modifications," each of which in turn was beneficial. After the sickle mutation first appeared, it began to increase in the population because of its beneficial effects, until millions of Africans had a copy of the gene. Now, instead of two mutations having to appear simultaneously in the DNA of one unbelievably lucky child, just one more mutation would have to happen in the offspring of any one of the millions of people who were already one step toward the goal. Because there were many more people playing the sickle Powerball lottery, all of whom had to match only one number instead of two, the jackpot went off relatively quickly.

Hemoglobin C-Harlem would be advantageous if it were widespread in Africa, but it isn't. It was discovered in a single family in the United States, where it doesn't offer any protection against malaria for the simple reason that malaria has been eradicated in North America. Natural selection, therefore, may not select the mutation, and it may easily disappear by happenstance if the members of the family don't have children, or if the family's children don't inherit a copy of the C-Harlem gene. It's well known to evolutionary biologists that the majority even of helpful mutations are lost by chance before they get an opportunity to spread in the population.[7] If that happens with C-Harlem, we may have to wait for another hundred million carriers of the sickle gene to be born before another new C-Harlem mutation arises.

Suppose, however, that the first mutation wasn't a net plus; it was harmful. Only when both mutations occurred together was it beneficial. Then on average a person born with the mutation would leave fewer offspring than otherwise. The mutation would not increase in the population, and evolution would have to skip a step for it to take hold, because nature would need both necessary mutations at once. For frail old men, the missing step would be a prohibitive barrier. The Darwinian magic works well only when intermediate steps are each better ("more fit") than preceding steps, so that the mutant gene increases in number in the population as natural selection favors the offspring of people who have it. Yet its usefulness quickly declines when intermediate steps are worse than earlier steps, and it is pretty much worthless if several required intervening steps aren't improvements.

Smith and Orr's prohibition of double mutations may be wrong for malaria, but it is right for species like humans and other large animals. Our measly population size of millions to billions puts us in the "frail old man" class. *Plasmodium falciparum,* however, with a yearly population size on the order of a hundred billion billion (10^{20}) or so is in the "couch potato" class, and can jump a missing stair or two.

COHERENCE

The second basic criterion for distinguishing between random and nonrandom mutation is *coherence.* Darwinian evolution cannot pursue a future goal. So envisioning Darwinian evolution as akin to climbing a solitary staircase—even one with missing steps—risks a subtle, yet fatal misconception. It is all too easy to think of the top of the stairs as the target, and to focus exclusively on the path leading to it, ignoring all other possibilities. If a Darwinist visualizes steps leading to some biological feature, the temptation is to conclude the route would be easily traveled by unaided nature. However, as Coyne and Orr emphasized, we need to ask whether a process is not just theoretically possible, but also biologically reasonable. Because random mutation and natural selection have no goal, Darwinian evolution faces the huge problem of incoherence: Like a drunkard's walk, the next evolutionary step a population of organisms takes is very likely to be unconnected to the last step. The upshot is that *even if* a gradual route toward a complex structure exists—*even one with no missing steps*—if the route is lengthy enough, the likelihood of reaching it by random mutation is terrible.

To grasp the problem of incoherence, picture a huge castle—much bigger than Iacocca Hall, much taller than six floors. At the very top of the castle is a single small room with a balcony, from which, say, a hero can wave to adoring crowds below. Getting to the top requires navigating a maze. In the castle there are no coherent staircases, only single steps. The bottom of the castle is a large room with a hundred doorways. Behind each door is a single step that opens onto another room (a different room for each of the doorways), which again has a hundred doorways leading to a single step, leading to a new room with a hundred doorways, and so on. Frequently, however, the step up through a doorway leads to a room with no other doorways and no other steps—a dead end. Only one route of the many possible ones leads to the window. An elderly, blind knight who is retiring from service begins climbing at the bottom of the castle with one thought in mind—to limp up any step he comes across, hoping to reach the balcony at the summit and receive the adulation he deserves. Unfortunately, because he is elderly, once he goes up a particular step he can't go back down; he would stumble and injure himself. So if he reaches a dead end, he's stuck. Although in a storybook the knight would surely find his way somehow, Darwinism professedly has no use for fairy tales. Almost every knight in a Darwinian story should get stuck in a dead end, never to reach the summit.

The blind knight's quest to climb to the summit of a castle that has no coherent staircase—only disconnected steps—mirrors a staggering difficulty for Darwinism: *Even if* there is some gradual route to a distant pinnacle, it is not "biologically reasonable" to expect random mutation and natural selection to navigate a maze to get there. Because steps are not preorganized into a staircase, and because so many wrong turns and dead ends lie in wait, an unseeing search would almost certainly fail. Without a floor plan or a guide, the knight would languish in some lower windowless room.

RUGGED LANDSCAPES

A problem akin to the knight's predicament has been discussed fitfully with no resolution in evolutionary biology journals under the name "rugged fitness landscapes."[8] In the 1930s the mathematical biologist Ronald Fisher pictured evolution as an exercise in hill climbing. The idea is that a species would gradually evolve to get better and better—to become more "fit"—until it was as good as it could be under the circumstances. In a sense, the species would rise to the acme of an evolutionary hill. Once there, it would be stuck—going back down the hill means getting less fit, which in a Darwinian competition should almost always be prohibited.[9] Well, what if, more realistically, instead of a single hill, the evolutionary geography actually resembled a badlands: a whole rugged landscape filled with many hills—big ones, little ones, tiny ones? The tiny ones are by far the most common, bigger ones much less frequent. There is only one highest peak. If so, then in a rugged evolutionary landscape, it is much more likely that a species will climb a tiny hill and get stuck there, unable to become less fit, yet forever isolated from the surrounding peaks. Random mutation and natural selection can't solve the rugged landscape dilemma—they actually *cause* the dilemma.

Even in the shadow of an evolutionary Mount Everest—the promise of some terrific new biological feature—the challenge of a rugged landscape would remain. In fact, that is where it would become especially difficult. The more complex and interactive a system, the more its simple variations will short-circuit evolutionary hill climbing. As a physical example, think of the goal of building a structure like Iacocca Hall. An evolutionary story might start with a small shack, useful as a shelter, and hope to build on that. But the materials one would use to build a shack (wood, straw, nails) are not the ones one would need for a larger structure (cement, steel). The shack would serve, for a while, but could not be altered into a large building without essentially being replaced.

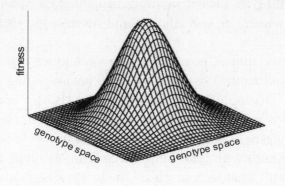

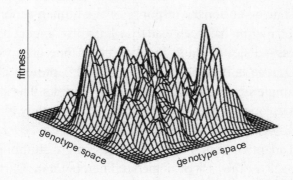

FIGURE 6.1

Evolutionary fitness landscapes. The top figure represents a simplistic evolutionary landscape, where only one or a few traits can vary, and fitness can increase smoothly. Ordinary Darwinian processes would easily drive a species to the single pinnacle. The bottom figure represents a more realistic, rugged evolutionary landscape, where many traits can vary. Here random mutation and natural selection would drive a species to some local peak, where it would remain stuck. Natural selection would actively inhibit a species from traversing such a landscape. If a limited scientific study focuses on just one peak of a rugged landscape, the results can misleadingly seem to match the smooth peak. (Reprinted from Gavrilets, S. 2004. *Fitness landscapes and the origin of species.* Princeton, N.J., Princeton University Press. Courtesy of Sergey Gavrilets.)

Yet tearing down the building would remove the only shelter available at the time. Even construction of a small building that improbably used cement and steel would not include spaces for future staircases, electrical wiring, and so on that would be needed for a larger building. A smaller building that did have space for them would very likely be less efficient and more costly than one that didn't.

To mix metaphors, how many steps should we expect random mutation and natural selection to climb before getting stuck on a tiny hill of a rugged landscape? Very few. Using a sophisticated mathematical model, H. Allen Orr decided that the likeliest number for a single gene was between just one and two.[10] That count fits pretty well both with John Maynard Smith's reasoning about proteins and with what we know from the best relevant data on evolution we have available—the effects of malaria on the human genome. The evolutionary response of the human genome to *Plasmodium falciparum* has been exactly what you'd expect of a Darwinian process—disjointed and incoherent. In one group of humans the G6PD gene is broken, in another band 3 protein is defective. Both are single steps to small, local adaptive peaks. The sickle mutation pops up once or a few times, and then, separately, alterations in fetal hemoglobin ameliorate its side effects—several steps to an unrelated adaptive peak. Like some blind knight stumbling through a castle maze, in the case of sickle/fetal hemoglobin, Darwinism has managed to walk up two steps, but has become stuck in an evolutionary dead end. Random mutation and natural selection are operating at full steam, but they lead nowhere.

This is not the kind of process that could have coordinated the many proteins that work in concert in intraflagellar transport. It is not the kind of process that leads to *any* significant degree of coherence.

AN ANALOGY OF INCOHERENCE

To get a better feel for the helplessness of Darwinism in the face of the problem of incoherence, let's return to another analogy. Suppose you enter a large room to find those proverbial million monkeys chained to keyboards. (Unlike the unruly, real-life macaques, these hypothetical fellows politely keep pecking away at all the keys.) They are hard at work revising *Moby Dick*. Because they are thoroughly modern monkeys who use word-processing computers instead of typewriters, not only can they change a letter here or there, they can also randomly delete or duplicate or rearrange entire passages, as well as add text, find and replace, and so on. In fact, since the computers are networked, sometimes the text on one computer can even randomly recombine with the text on another computer. These extra abilities nicely mimic the variety of ways in which DNA can mutate in the cell.

Altered texts are offered for sale to the public, and the sales of a version determines its fate. If a change is for the worse, as most would be, then it sells poorly. If it sells less well than any other copy, the marvelous computer erases it and reverts to the original text on the computer before the monkey made the most recent alteration. However, if a revision improves the text—so that it would sell more copies—then it somehow becomes the standard text against which changes are judged not only in the computer of the monkey who originally typed it, but also in some of the other monkeys' computers.

What would count as an improvement? A variety of changes. One way to improve *Moby Dick* would be to make the novel easier to read. Another way would be to make it more entertaining. Another to make it more profound. Another to make the book cost less. The book might cost less if it were shortened. It might become more profound by repeating profound passages, or by including profound passages from other texts (like, say, the Gettysburg Address), which may also be stored in the computer. To be more entertaining it might contain silly misspellings, or jokes copied from other texts. To be easier to read it might delete words, sentences, or much more (à la *Reader's Digest*). In short, there are many, many ways in which the complex text might change to improve the book. In fact, there are so many possible improvements that we should not expect changes to be substantially connected to each other. They are likely to accumulate independently and incoherently.

To illustrate, let's look at some of them. One of the earliest monkey alterations is the insertion of a comma into the first sentence, changing it from "Call me Ishmael" to "Call me, Ishmael." But that doesn't help at all. It doesn't make the book more profound or cheaper. It makes the book more difficult to read, since the first sentence now fits poorly with the rest of the book. The problem of coherence increases as the complexity of a text increases. If "Call me Ishmael" were the whole text, then inserting a comma after "me" would significantly change its meaning without conflicting with the meaning of later text. But since it's part of a longer, coherent story, it doesn't fit. The book doesn't sell, so the alteration is erased and the text reset to the original.

By contrast, another early change, a duplication, is successful. In the final chase for the white whale in the last chapter, in one sentence "even now," is repeated several times to give "'Oh! Ahab,' cried Starbuck, 'not too late is it, even now, even now, the third day, to desist.'" The change strikes some readers as more dramatic and profound, and thus spreads to some other monkey computers. In another alteration the second, descriptive paragraph of Chapter 44 is deleted.[11] The change interrupts the narrative flow but, because it also shortens the text slightly, it makes it a bit more readable; on balance it is a slight improvement. It spreads to some other computers. A third beneficial change is the deletion of Chapter 6, mostly description. A fourth change is the duplication of the epilogue, which counts as more profound (don't ask blind Nature to act as a picky literary critic, please). A fifth switch substitutes "Arab" for "Ahab" throughout the story. A sixth recombines the text on a computer missing Chapter 6 with one that has a duplicated epilogue. In between all these major changes are many smaller ones where individual words and phrases are deleted, rearranged, misspelled, duplicated, and so on. Occasionally, a change might build upon a previous change—maybe successive paragraphs might be deleted in two separate steps, or maybe several alterations change, say, WORD to WORE to GORE to GONE to GENE, and perhaps that might help somehow—but those alterations are no more likely to be helpful than many other possible improvements.

In the end, although many changes accrue to the text, and even though the text is in a sense "improved" in that it sells better than the original edition, the changes do not add up to anything like a coherent new story. There is no new ending where, say, Ahab survives and sells the blubber of Moby Dick for a fortune, or where Ishmael recounts his earlier life before going to sea. Writing a coherent story of course requires an author like Herman Melville, who can visualize the storyline in its overarching complexity.

How many changes could be made randomly to a relatively small piece of text—a sentence or paragraph—before it comes to a local dead end, better than all other single changes surrounding it, but still not very much improved from the starting text? Although he was considering biological texts, Allen Orr's work on genes is likely applicable here as well. If so, we should expect the answer to be one or two changes, possibly several. A deletion or two in a sentence or paragraph might make the text more readable, but more deletions might interfere with the sense of the text. The same with other changes—rearrangements, substitutions, and so on. As with smaller pieces of text such as words and sentences, so, too, with larger ones like book sections and chapters. Switching or deleting a chapter or two in a book might make it more readable; further large changes likely wouldn't.

The eminent geneticist François Jacob famously wrote that Darwinian evolution is a "tinkerer," not an engineer.[12] He's exactly right. Tinkering means looking for quick fixes, features that work for the moment—incoherent, patchwork change, doctoring machines with chewing gum and duct tape, stopping an invader by burning a bridge or breaking a lock, "improving" a text by typing disjointed changes to words, letters, paragraphs, and chapters. If Darwinism is a tinkerer, then it cannot be expected to produce coherent features where a number of separate parts act together for a clear purpose, involving more than several components. Even if someone could envision some long, convoluted, gradual route to such complexity, it is not biologically reasonable to suppose random mutation traversed it. The more coherent the system, and the more parts it contains, the more profound the problem becomes.

A MINOR FACTOR

The degree of coherence of a system and the number of steps that have to be skipped to get from one level to another are the two major criteria by which we can try to locate the edge of evolution. There is also a lesser factor that sometimes has to be taken into account to minimize our chances of being misled: degradation—when a more complex system breaks down to yield less complex systems. For example, suppose an automobile fell apart. We might be able to salvage from it a number of separate parts, such as a radio, air conditioner, or pump. Or consider that one might come across the isolated phrase "Call me Ishmael" and not have any idea that it derived from a larger system. In deciding how a particular feature arose, we have to consider whether it had originally been a piece of a larger system. Of course, when degradation does occur, it simply means that the question of Darwinian randomness must be readdressed to the preceding structure.

COHERENCE, STEPS, AND IRREDUCIBLE COMPLEXITY

How does irreducible complexity, which I first described in *Darwin's Black Box* and discuss in the last chapter, fit with the criteria of coherence and evolutionary steps introduced in this chapter? Although closely related to it, the new concepts gauge difficulties for random mutation at a much finer level than does irreducible complexity, and are more appropriate to defining the edge of evolution. If irreducible complexity is likened to a rough measuring tool—say, a yardstick with no markings—then the new criteria can be thought of as a ruler subdivided into millimeters.

Let's reconsider the mousetrap. A common mechanical mousetrap is an example of irreducible complexity because it "is composed of several well-matched, interacting parts that contribute to the basic function" and "the removal of any one of the parts causes the system to effectively cease functioning."[13] The mousetrap is resistant to gradual Darwinian-style explanations.

Yet a boatload of unnoticed action is packed into the terms "well-matched" and "interacting." The spring of a mousetrap isn't just any old spring. For example, the spring from a grandfather clock or windup watch or toy car would be useless in a standard mousetrap. In fact, the spring in a mousetrap is essentially unique. Its length is critical to its role in the trap. If there were somewhat fewer or more coils the ends of the spring would not be positioned correctly. The ends of the spring themselves are not coiled; rather, they're extended. One end presses against the wooden platform and the other oddly shaped end hooks over the hammer—the part of the trap that strikes the mouse. The positioning of both ends is crucial to the work of the spring. If either end were at a somewhat different angle coming off the main body of the spring, the trap would be ineffective. If either end were substantially longer or shorter, the system would again be compromised. It's clear that not only the whole trap, but also the spring and other pieces required multiple, coherent steps to produce.

At the risk of appearing pedantic, let me list some steps. Suppose that a smith intends to make a spring for a mousetrap. To do so he takes a series of actions. In his shop he has a large number of lengths of spring of various shapes and sizes that he keeps in store for projects. Out of the hundreds of stock springs he chooses one with the right diameter and resilience. He cuts a one-and-three-quarters-inch length off the end of the yard-long stock. He heats up one end in a flame and straightens out the metal to a length of one inch. He does the same to the other end. He positions one end at an angle of 180 degrees with respect to the other end. Then he puts a crimp in the end that will overlap the hammer when the trap is put together. The lesson is obvious: Simply to fashion the spring into the correct special shape needed for the trap requires multiple, coordinated steps.

The concept of irreducible complexity, with its broad focus on the "parts" of a system, passes over the fact that a part might itself be a special piece that needs explaining in terms of many steps. What's more, it also overlooks the steps required to assemble a system—to physically put them together—once the parts are available. Once the spring is forged and the other pieces of the future mousetrap manufactured, the smith grabs the various pieces lying at different spots in the shop, transports them to his workbench, and pieces them together in the right orientation. The concept of the number of "steps" resembles the idea of irreducible complexity in that both look to see if multiple factors are needed to produce something. But "steps" goes further, asking how many separate *actions*—not just separate *parts*—are needed to make a system. The concept of "steps" is especially useful when fewer actions are needed to coherently arrange parts. It can locate the edge of evolution with greater precision.

The concept of coherence is implicit in the definition of irreducible complexity in the idea of parts that are "well matched" to a "system." The standard mechanical mousetrap, with its very well-matched parts, is profoundly coherent. Since it is irreducibly complex, it can't be built directly by a gradual process that would mimic a Darwinian scenario. But suppose there were some tortuous, indirect route that might lead to the trap. It would be unreasonable to expect the route to be found by a blind process, for the same reason that we wouldn't expect the blind knight to maneuver through a maze to the summit of the castle—there are too many dead ends and opportunities to go wrong. To trap mice, a deep hole in the ground might do just fine. Yet a hole in the ground isn't a route to the standard mechanical mousetrap. If the hole then had to be filled in before starting over to build a better mousetrap (pardon the strained analogy), then mice would flourish at least temporarily—ruling out this path. A splotch of glue can catch a mouse, but can't be turned into a mechanical trap. If the glue trap had to be discarded before starting over to make a mechanical trap, we'd be worse off than before.

The more pieces, and the more intricately they interact, the more opportunities there are to go wrong in building a system. Even with small systems, one can go wrong right off the bat, get trapped in a dead end, and never make it to the top of the castle. If Herman Melville had started writing his novel with "Call me, Ishmael," *Moby Dick* would have gotten off on the wrong foot and might never have been written. The benchmarks discussed in this chapter are thus a better guide to the edge of evolution than is irreducible complexity. Now, it's time to get down to business, using these benchmarks to try to define the ragged edge of Darwinian evolution.

7

THE TWO-BINDING-SITES RULE

A GOOD FIT IS HARD TO FIND

In *Darwin's Black Box* I explained how design can be apprehended in the arrangement of parts. It can also be perceived in the *fitting* of complex parts, even if the reason for the fitting is obscure. Here's one example. Suppose that you were walking through a junkyard. All sorts of complex parts were lying around: a pipe to your left, springs to your right, bolts, screws, pieces of metal, and much more. Although the yard was filled with many manufactured parts, there would be no reason to think they had anything to do with each other. However, suppose you spotted a compact pile of parts. When you picked up one of the parts, the rest came along—they were attached. What's more, you saw that the parts matched closely—holes in one piece were aligned to pegs in another; curves fit with indentations, and so on. Even if you didn't know the function of the aggregate of pieces, you would be pretty sure they had been put together on purpose, unlike the other parts in the junkyard, because they specifically fit each other.

In order to be sure that parts are designed to fit each other, their shapes must be relatively complex and must match each other pretty closely. If the shapes of pieces are comparatively simple, or if the fit is quite loose, their complementarity may just be a matter of luck. For example, it's hard to decide if, say, a generic book was intended to be mailed in a generic box. The rectangular shapes of both a book and a box are pretty simple, so even if the book fits loosely in the box, it might be a coincidence. On the other hand, it's easier to conclude that a box with exactly the right-sized compartments was built to ship a particular computer and accessories if all the pieces fit snugly together, with projections and indentations of the computer nicely accommodated by complementary indentations and projections of the box (or, say, Styrofoam packing material). The greater complexity of the second box allows for a firmer decision of purposeful design.

In judging whether two unfamiliar parts were designed to fit each other, one has to be careful. The likelihood of finding two parts that fit by accident has to be weighed against the number of different parts that are on hand. If a warehouse were filled with a million small, rigid, plastic pieces of all different shapes, it wouldn't be too surprising to find one that, say, fit pretty well inside a heart-shaped locket. The more and more pieces we have on hand, the more and more likely we are to find two that fit each other closely, just by chance. On the other hand, the more and more complex the shapes, the less likely.

PROTEIN PIECES

Proteins have complex shapes, and proteins must fit specifically with other proteins to make the molecular machinery of the cell. Although most proteins were once thought to act individually, in the past few decades science has unexpectedly discovered that most proteins in the cell actually work as teams of a half dozen or more. As former president of the National Academy of Sciences Bruce Alberts remarked:

> We can walk and we can talk because the chemistry that makes life possible is much more elaborate and sophisticated than anything we students had ever considered. . . . [I]nstead of a cell dominated by

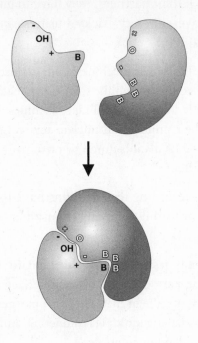

FIGURE 7.1

Cartoon of a smaller protein binding to a larger one. Both the shapes and the chemical properties of the protein surfaces must be complementary to bind. (+ and - stand for positively and negatively charged groups on the protein surfaces. B stands for hydrophobic, oily groups. OH and O stand for polar groups that can "hydrogen bond" to each other.)

randomly colliding individual protein molecules, we now know that nearly every major process in a cell is carried out by assemblies of 10 or more protein molecules. And, as it carries out its biological functions, each of these protein assemblies interacts with several other large complexes of proteins. Indeed, the entire cell can be viewed as a factory that contains an elaborate network of interlocking assembly lines, each of which is composed of a set of large protein machines.[1]

What's more, in contrast to the machines in our everyday world, proteins must *self-assemble*. Although machine parts in our familiar world must have complementary shapes, they are put together by people (or robot surrogates). Protein parts in cellular machines not only have to match their partners, they have to go much further and assemble themselves—a very tricky business indeed. As a recent issue of *Nature* put it:

> The cell's macromolecular machines contain dozens or even hundreds of components. But unlike man made machines, which are built on assembly lines, these cellular machines assemble spontaneously from their protein and nucleic-acid components. It is as though cars could be manufactured by merely tumbling their parts onto the factory floor.[2]

To perform that astounding feat, proteins have to pick their correct binding partners out from the many thousands of other proteins in the cell:

> A protein generally resides in a crowded environment with many potential binding partners with different surface properties. Most proteins are very specific in their choice of partner, although some are multispecific, having multiple (competing) binding partners on coinciding or overlapping interfaces.[3]

In order to assemble correctly, the absolute minimum requirement is that proteins must stick specifically to their partners in the right orientation. As shown in Figure 7.1, not only do the shapes of two proteins have to match, but the chemical properties of their surfaces must be complementary as well, to attract each other. If the shapes of two protein surfaces match each other but their chemical properties don't, the two surfaces won't stick; they might bump together in the cell, but if so they would quickly drift apart.

Were complex protein-protein interactions designed, like so many complementary automobile parts? Or are they merely accidental, more like the heart-shaped piece from a large warehouse that fits into a locket? Or are there some in each category—some interactions arising by serendipity, others through intent? Over the next several sections we'll look closely at how proteins choose a partner and consider the large obstacles that the evolution of new protein-protein interactions presents to random mutation. We'll see that it is reasonable to conclude that, although some interactions are accidental, the great majority of functional protein interactions arose nonrandomly. And that means nonrandomness extends very deeply into the cell.

BINDING EVERYTHING

Much has been learned about how proteins bind to each other by studying the immune system. The complex vertebrate immune system protects against invasion by microscopic predators. One facet of the system generates a prodigious number of different proteins called antibodies that patrol the circulatory system. Although much of their structure is pretty similar, the antibodies differ from one another at one end, called the binding site. If a foreign cell or virus is lurking in the circulatory system, one or more antibodies will likely stick to the invaders via the binding site, marking them for destruction by other immune system components.

In order for an antibody to stick, the binding site must be geometrically and chemically complementary to the foreign surface. In early work on the immune system, before it was realized just how clever the system is, it was natural to think that vertebrate antibodies had been shaped by natural selection to recognize the surfaces of previously encountered pathogens, so that the only antibodies expected to be available would be ones that had binding sites to past or present invaders. However, subsequent experiments showed that when test animals were injected with synthetic chemicals that had likely never before existed on earth, some existing antibodies were able to bind to the manmade materials. That meant that some binding sites were complementary to shapes that the animal or its ancestors had never encountered! How could that be?

A number of ideas were floated that turned out to be wrong. One proposal, advanced by the two-time Nobel Prize winner Linus Pauling, was that antibody binding sites "mold" themselves to the foreign chemical—wrap around it and "freeze" in position.[4] Another was that perhaps the vertebrate genome carried genes for a huge number of antibodies, and by luck the shapes of some of their binding sites matched the shapes of molecules that hadn't even been invented by nature. That was closer to the mark, but it greatly underestimated the elegance of the design of the immune system. It eventually became clear that, while vertebrates do make a huge number of different antibodies (many billions), they have a limited number of antibody genes (a few hundred). Much hard lab work eventually showed that the trick to generating many different antibodies from a small set of genes is the same principle that allows a huge number of different poker hands to be dealt from a deck of just fifty-two cards. In brief, immune cells contain specific molecular machinery that shuffles segments of genes (and performs other tricks, too), allowing very many antibodies to be produced with an extraordinary diversity of binding sites.

SHAPE SPACE

Over the past decades immunologists have injected test animals with a wide range of synthetic chemicals, and almost unfailingly the animals prove to have antibodies that counteract them. To explain the ability of the immune system to do this, the mathematician Alan Perelson invented the concept of "shape space."[5] Shape space envisions a sort of library of physical objects roughly the size of a protein-protein binding site, with indentations and projections, smooth surfaces and rough ones. Perelson calculated that the huge number of different antibody binding sites made by the immune system was enough to contain essentially every possible shape, so that there would be at least one antibody to bind reasonably well to proteins on the surface of an invading bacterium or virus, even if the body had never been exposed to it before. To revert to my analogy of a warehouse full of plastic shapes, this shows that the binding matches between some antibody and some synthetic chemical were *not* specifically designed (although the clever, special mechanisms to enable the immune system to produce all those antibodies very likely were). The universe of antibodies is for all practical purposes infinite, which precludes an inference of design for these instances of binding. The knowledge gained from studying the immune system is invaluable for assessing the edge of evolution elsewhere. Is evolution full of processes analogous to our immune system, with automatic shufflings that make all things possible?

Like something from a science fiction story, the objects in Perelson's protein shape space are multidimensional—they have more than the usual three dimensions found in our ordinary world. They also have "dimensions" that take into account their varying chemical properties—positive charges, oily areas, and more. Imagine a huge library of complexly shaped objects that also had a half dozen or so weak bar magnets inserted at irregular spots, with just the tip of each magnet sticking out on the surface. The magnets could be oriented with their north poles facing either in or out, so that the magnet tips on the surface could either attract or repel each other. The extra dimensions of shape space can be thought of as accounting for the placement and orientation of the magnets. So shape space accounts not only for the physical shapes of objects, but also for their ability to stick to each other.

Suppose that thousands of objects from the shape space library are placed in a well-stirred swimming pool, and a prime goal for each is to find its one ideal mate. Reaching that goal faces a "Goldilocks" problem. First, consider objects that bind indiscriminately. Suppose some shape-space objects weren't very rigid—they had flexible octopus arms lined with magnets that allowed them to stick to many other objects in the swimming pool. Although they'd bind strongly, those stick-to-everything objects would gum up the works. The cell cannot tolerate objects that bind haphazardly.[6] They must be eliminated. Next, think about objects that are rigid, but don't match. Suppose a pair of objects do have complementary shapes, but they don't have magnets lined up in the right positions. Although, by rotating around, one or two pairs of magnets might be brought close to each other, that's not nearly enough to hold them together in the swirling waters of the pool, so they immediately fall apart. The two lessons are: 1) Nonspecific, octopuslike objects that bind strongly to many other objects are hazardous and must be removed; and 2) in order to stick specifically and well, two rigid objects have to fit each other in both shape and magnet pattern.

Binding doesn't have to be all or nothing. A pair of rigid objects might have shapes that are pretty complementary, and magnets that

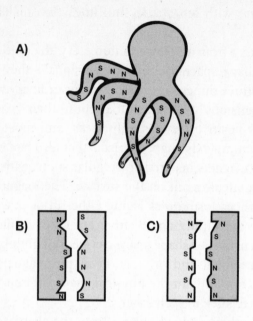

FIGURE 7.2

Cartoon of objects from a shape-space library. N and S stand for the north and south poles of magnets. (A) The floppy, octopuslike shape would bind non-specifically to many objects. (B) These rigid objects are complementary in magnet pattern, but not in shape, so they would not bind well to each other. (C) These rigid objects are complementary in shape, but not in magnet pattern, so they would not bind well to each other either.

are fairly well aligned. They might stick pretty well, but every so often a strong current in the pool or unusually big collision knocks them apart. After stirring for a while they reconnect. On average they might spend about half their time together and half apart. On the other hand, another pair of objects might fit together like a hand in a glove, with all magnets aligned perfectly. They might stick so tightly that they rarely fall apart and spend about 99 percent of their time together.

HOW BIG IS SHAPE SPACE?

We can adapt the lessons of the immune system and shape space to help understand the problems random mutation would face in making new protein-protein binding sites in the cell. The immune system is set up to get around the problem. But what happens when you remove the manufacturer who threw all these shuffled pieces of plastic into the pool? What if new pieces had to be made by rare dents or scrapes to old pieces? The immune system is capable of producing an essentially infinite warehouse of little plastic shapes. Experiments with similar systems in laboratories have enabled scientists to test finite subsets of that warehouse, to see how big it must be in order to handle particular challenges. These experiments yield generalizable parameters for the limits of binding proteins in any context, not just immune systems.

A huge hurdle confronting Darwinian evolution is the following: Most proteins in the cell operate as specific complexes of a half dozen or more chains.[7] Hemoglobin comprises a complex of two kinds of amino acid chains (alpha and beta) stuck together, but hemoglobin is relatively simple. Most cellular proteins have six or more kinds of amino acid chains. So, unless those complexes were all together from the start, then at some time in the past separate cellular proteins had to develop the ability to bind to each other. But that would be a very tricky business indeed. On the one hand if, like the octopus objects above, a protein developed a surface that stuck indiscriminately to a lot of other proteins, it would gum up the workings of the cell. It would have to be eliminated. On the other hand, most protein pairs wouldn't bind to each other at all, or bind very weakly, because their surfaces don't match closely enough. Only when a Goldilocks match randomly developed between their multidimensional surfaces would two proteins bind to each other tightly and specifically enough to make an effective pair. So we can ask, how difficult would it be for two proteins that initially did not bind to each other to develop a strong, specific interaction by random mutation and natural selection?

To start to answer that question, let's take the measure of shape space. Is shape space small, medium, or huge? How many protein binding sites do we need in our shape-space library to find a decent match for another given protein? In the 1990s Greg Winter and coworkers at the Medical Research Council in Cambridge, England, were interested in developing artificial antibodies to use as tools in medical research and treatment. In a series of papers they focused on the question of how large a shape-space library would be needed to have a good chance of containing at least one antibody with a binding site that would stick pretty specifically to an arbitrary test protein. Using clever laboratory methods, they made antibodies in which the amino acids of the binding site had been randomized, and generated a very large number of different combinations—a hundred million, in fact.

Then they went fishing. They used various molecules—either a different protein or other chemical—as bait and tried to pull out from the mixture of antibodies ones that would bind to the bait. They saw that, like the pieces in the swimming pool that would stick to each other about half of the time, on average the antibodies they isolated could bind to the bait with just moderate strength.[8] Over the years they and other laboratories juiced up the size of the antibody library, from a hundred million to a hundred billion and more. They found that the strength of binding improves with the size of the library, so that the best binders from the big library stick very tightly to the bait,[9] like the pieces in the swimming pool that spent 99 percent of their time together. This result—the bigger the library the better the binding—is pretty much what you would intuitively expect. After all, if by chance you find a shape in a smaller library that binds something reasonably well, then you have a chance with a larger library to find a shape that fits even better.

The general results from Winter's lab have been consistently confirmed: In order to get a particular protein to bind to any other one with modest strength, on average you have to wade through about ten to a hundred million binding sites.[10] Actually, these and other experimental results are strongly skewed in a way that underestimates library sizes that would be needed if mutation were truly random. In all of these experiments, mutations were deliberately confined to a coherent patch of amino acids that were close to each other on the surface of the protein, to make as many novel, binding-site-sized regions as possible. If the workers had not deliberately directed the changes to a coherent patch on the protein's surface, most changes would be scattered, unable to effectively interact. In that case a very much larger number of mutations on average would have to be sifted to find one that stuck specifically to a target protein. So we can take the results of these experiments as a very optimistic estimate for the difficulty of searching shape space.

LOST IN SHAPE SPACE

The elegant immune system is designed to saturate shape space. But the situation is entirely different inside the cell. For cellular proteins there is no built-in mechanism to deliberately make new binding sites. Cellular proteins almost always are made with just one sequence, not billions of different sequences like antibodies. In general the only way to get a new sequence for a cellular protein is over many generations by random mutation. Searching through shape space with cellular proteins is glacially slow and abysmally inefficient.

How much of a hurdle is it for Darwinian evolution? Consider a hypothetical case where it would give an organism some advantage if a particular two of its proteins, which had been working separately, bound specifically to each other. Perhaps the two-protein complex would be able to perform some new task, or do an old task much better. The lesson from shape space is that, in order for the one to bind the other, we should expect to have to search through tens of millions of different mutant sequences before luckily happening upon one that would specifically stick with even modest strength, which would allow the two to spend even half of their time together. (This is likely the minimum necessary strength, enough to have a noticeable biological effect.)[11] Since the mutation rate is so low—about one mutation at a particular site in a hundred million births—we would expect to have to slog through an enormous number of organisms before striking on that lucky one.

Let's make a rough calculation for the average number of organisms we would have to slog through to find a new protein-protein binding site. As I said, shape space tells us that about one in ten to a hundred million coherent protein-binding sites must be sifted before finding one that binds specifically and firmly to a given target. The simplest way to alter a protein is by point mutation, where one amino acid is substituted for another at a position in a protein.

There are twenty different kinds of amino acids found in proteins. That means that if just five or six positions changed to the right residues—the ones that would allow the two proteins to bind—that would be an event of approximately the right frequency, since twenty multiplied by itself five or six times (20^5 or 20^6) is about three million or sixty million, respectively—relatively close to the ten to a hundred million different sites we need.

So one way to get a new binding site would be to change just five or six amino acids in a coherent patch in the right way.[12] This very rough estimation fits nicely with studies that have been done on protein structure.[13] Five or six amino acids may not sound like very much at first, since proteins are often made of hundreds of amino acids. But five or six amino acid substitutions means that reaching the goal requires *five or six coherent mutational steps*—just to get two proteins to bind to each other. As we saw in the last chapter, even *one* missing step makes the job much much tougher for Darwin than when steps are continuous. If multiple steps are missing, the job becomes exponentially more difficult.

Let's consider one further wrinkle. Most amino acid changes in proteins diminish a protein's function. But about one-third of possible amino acid changes are like switching a k for a c in "cat" or "candy"; they can be accommodated without too much trouble.[14] Such "neutral" changes can occur during evolution and spread around a population by chance. So let's suppose that of the five or six changes that have to happen to a protein to make a new binding site, a third of them are neutral. They could occur before the other key mutations, as a separate step, without harm. Although finding the right neutral changes would itself be an improbable step, we'll again err on the conservative side and discount the average number of neutral mutations from the average number of total necessary changes. That leaves three or four amino acid changes that might cause trouble if they occur singly. For the Darwinian step in question, they must occur together. Three or four simultaneous amino acid mutations is like skipping two or three steps on an evolutionary staircase.

Although two or three missing steps doesn't sound like much, that's one or two more Darwinian jumps than were required to get chloroquine resistance in malaria. In Chapter 3 I dubbed that level a "CCC," a "chloroquine-complexity cluster," and showed that its odds were 1 in 10^{20} births. In other words (keeping in mind the roughness of the calculation):

> *Generating a single new cellular protein-protein binding site is of the same order of difficulty or worse than the development of chloroquine resistance in the malarial parasite.*

Now suppose that, in order to acquire some new, useful property, not just one but *two* new protein-binding sites had to develop. A CCC requires, on average, 10^{20}, a hundred billlion billion, organisms—more than the number of mammals that has ever existed on earth. So if other things were equal, the likelihood of getting two new binding sites would be what we called in Chapter 3 a "double CCC"—the square of a CCC, or one in ten to the fortieth power. Since that's more cells than likely have ever existed on earth, such an event would not be expected to have happened by Darwinian processes in the history of the world. Admittedly, statistics are all about averages, so some freak event like this *might* happen—it's not ruled out by force of logic. But it is not biologically reasonable to expect it, or less likely events that occurred in the common descent of life on earth. In short, complexes of just three or more different proteins are beyond the edge of evolution. They are lost in shape space.

And the great majority of proteins in the cell work in complexes of *six* or more. Far beyond that edge.

TOUCHING BASE

In science as in other areas of life, it's easy to fool yourself if you aren't careful. A lot of ideas seem plausible at first blush, but when you check them against the facts they don't work out. Reasoning in the abstract about shape space and what that implies for Darwinian evolution is a good first step, but how does it square with the data? In the next few sections we'll survey what we know from the best sources of evolutionary data available.

In its battle with malaria the human genome has been terribly scarred. In the past ten thousand years a number of genes have been broken or their efficiency reduced in order to fend off malaria (as discussed in Chapter 2). Other than sickle hemoglobin (an exception we'll discuss in the next chapter), has the war with malaria caused humanity to evolve any new cellular protein-protein interactions? No. A survey of all known human evolutionary responses to the parasite includes *no* novel protein interactions. Although it can't be ruled out that some such thing has developed but escaped detection, we can be certain that its effects are (or were) weaker than those of the sickle mutation, thalassemia, and the other simple fractured genes, because they did not prevail over time.

Since malaria first appeared in its most virulent form about a hundred centuries ago, more than a billion humans have been born in infested regions. So, although it's risky to draw too firm a conclusion from just one example, it appears that the likelihood of the development of a new, useful, specific protein-protein interaction is less than one in a billion organisms.

Conversely, in its battle with poison-wielding humans, the malaria genome has also been terribly scarred. In the past half century a number of genes have been broken or altered to fend off drugs such as chloroquine. As discussed in Chapter 3, none of the changes seem to be improvements in an absolute sense. They disappear once drug therapy is discontinued. Has the war with humanity caused malaria to evolve any new cellular protein-protein interactions? No. A survey of all known malarial evolutionary responses to human drugs includes *no* novel protein-protein interactions. Although, as above, it can't be ruled out that some such thing developed in the past, no such change persisted, so none could have been as effective as the damaging changes discussed earlier.

Since widespread drug treatments first appeared about fifty years ago, more than 10^{20}, a hundred billion billion, malarial cells have been born in infested regions. It thus appears that the likelihood of the development of a new, useful, specific protein-protein interaction is less than one in 10^{20}. Since sickle hemoglobin, thalassemia, and other human genetic responses have appeared, probably another thousandfold *P. falciparum*, 10^{23}, have infected humans, with no known protein-protein interactions, or any other effective response, having developed. So it seems that the odds of the development of a new, useful, specific, protein-protein interaction are less than one in 10^{23}—worse than a CCC.

AIDS AND EVOLUTION

Studies of malaria provide our best data about what Darwinian evolution can do, but there are other studies of interest. One excellent source of information comes from the study of the human immunodeficiency virus HIV, the virus that causes AIDS. Like malaria, HIV is a well-studied scourge and killer that first appeared in Africa. Unlike malaria, HIV is spread by person-to-person contact, so it can survive in mild and even cold climates. Also unlike P. falciparum—which is a eukaryote, the most complex type of cell—HIV is a virus, one of the simplest forms of life. The amount of genetic information in the AIDS virus is less than a thousandth the amount of DNA in the malarial parasite. What's more, viruses such as HIV mutate much more readily than cells do—about ten thousand times faster. The HIV virus is so small, and the mutation rate is so great, that on average each new copy of the virus contains one change, one mutation, from its parent. HIV mutates at the evolutionary speed limit—Darwinian evolution just can't go any faster.

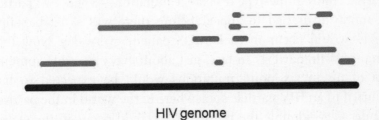

HIV genome

FIGURE 7.3
Schematic diagram of the genome of HIV. The black bar represents the intact genome. The gray bars show the approximate location of the nine viral genes in the genome. (Gray bars connected by a dashed line represent genes that are pieced together.) The virus is about one-millionth the size of the human genome. Its basic genetics have changed very little in the past decades, despite an enormous mutation rate and the production of a hundred billion billion copies.

About a hundred billion billion, 10^{20}, malarial cells are born each year. The best current estimate is that a person infected with HIV is burdened with a total of one to ten billion (10^9 to 10^{10}) virus particles.[15] The generation time for virus replication is about a day or two[16], so over the course of ten years a single person will produce more than a thousand generations of HIV, or up to 10^{13} viruses. Since there are approximately fifty million people worldwide infected with the virus, the math points to a total of about 10^{20} copies of the virus having been produced in the past several decades, when HIV became widespread in human populations— roughly the same as the number of malarial cells produced each year.

But the total number of copies of the virus is only part of the story. The other important factor is the speeded-up evolution of HIV due to its much greater mutation rate. Because of the difference in mutation rates HIV has actually experienced about ten thousand times as many mutations as would a comparable number of malarial cells. The very many copies of HIV in the world would be expected to contain almost every imaginable kind of mutation. As one study put it, "Each and every possible single-point mutation occurs between 10^4 and 10^5 times per day in an HIV-infected individual."[17] Every double point mutation, where two amino acids are changed simultaneously, would occur in each person once each day. (This means a chloroquine-type resistance mutation—where two particular amino acids had to appear before there was a net beneficial effect—would occur in each AIDS patient every day. Now that's mutational firepower!) In fact, just about every possible combination of up to six point mutations would be expected to have occurred in an HIV particle somewhere in the world in the past several decades—double the number that could occur in the slower-mutating *P. falciparum*. In addition to all those point mutations, enormous numbers of insertions, deletions, duplications, and other sorts of mutations would occur as well.

And exactly what has all that evolution of HIV wrought? *Very little.* Although news stories rightly emphasize the ability of HIV to quickly develop drug resistance, and although massive publicity makes HIV seem to the public to be an evolutionary powerhouse, on a functional biochemical level the virus has been a complete stick-in-the-mud. Over the years its DNA sequence has certainly changed. HIV has killed millions of people, fended off the human immune system, and become resistant to whatever drug humanity could throw at it. Yet through all that, there have been no significant basic biochemical changes in the virus at all.

With a few apparent exceptions,[18] HIV enters its target cells of the immune system by first binding tightly and specifically to one of the many kinds of proteins on their surface, and then reaching over to bind another protein called a coreceptor. (Some humans are resistant to HIV because they burn the bridge that the virus uses to invade the cell: They have a broken copy of the gene for a coreceptor.) A hundred billion billion mutant viruses later, HIV continues to do exactly the same thing, to bind the same way. If a mutant virus developed the ability to enter other kinds of cells by binding to other kinds of proteins, it might replicate more effectively and thus outcompete its siblings. That hasn't happened.[19] Neither has much else happened at a molecular level.[20] No new gizmos or basic machinery. There have been no reports of new viral protein-protein interactions developing in an infected cell due to mutations in HIV proteins.[21] No gene duplication has occurred leading to a new function. None of the fancy tricks that routinely figure in Darwinian speculations has apparently been of much use to HIV.

But what about its ability to quickly evolve drug resistance and evade the immune system? Doesn't that show that Darwinian evolution is very powerful? Isn't that a sophisticated maneuver? No. It turns out that HIV employs the same modest tricks that malaria uses to evade drugs—mostly simple point mutations to decrease the binding of the poison to its pathogen target. For example, a change of just one amino acid at position 184 of one particular HIV enzyme causes a little bump that interferes with one drug.[22] Another major drug target is a protein called HIV protease, which is a kind of special scissors needed to cut out some other viral proteins from their immature form. Typically, drugs are made that can stick to the protease and gum it up. And just as typically, point mutations appear that alter the protein shape a bit, so the poison doesn't stick so well.[23] Like the development of resistance to rat poison by rats, resistance of HIV to drugs is a very simple biochemical affair.[24]

Let's compare the results of HIV evolution to malaria evolution, and consider the changes both have wrought in humans. The number of copies of both in the last fifty years are roughly comparable—roughly, in the sense that malaria outnumbers HIV by a factor of only ten or a hundred. The number of genes in malaria is in the thousands; HIV has just nine. The mutation rate of HIV is greater than that of malaria by a factor of ten thousand. So the small HIV genome has been riddled by changes to a limited number of genes; malaria has endured a roughly comparable number of mutations, but spread out over a much larger genome. Nonetheless, despite the many differences between them, the evolutionary changes in both in the past fifty years are comparable and—despite their severe consequences for public health—biochemically trivial. A few point mutations, the occasional gene duplication in malaria; but no new, useful protein-protein interactions, no new molecular machines. The biochemical changes they have triggered in humans are comparable as well: a long list of broken genes for the ancient P. falciparum; a shorter list for the recent HIV.

The bottom line: Despite huge population numbers and intense selective pressure, microbes as disparate as malaria and HIV yield similar, minor, evolutionary responses. Darwinists have loudly celebrated studies of finch beaks, showing modest changes in the shapes and sizes of beaks over time, as the finches' food supplies changed. But here we have genetic studies over thousands upon thousands of generations, of trillions upon trillions of organisms, and little of biochemical significance to show for it.

LAB STAR

The studies of malaria and HIV provide by far the best direct evidence of what evolution can do. The reason is simple: numbers. The greater the number of organisms, the greater the chance that a lucky mutation will come along, to be grabbed by natural selection. But other results with other organisms can help us find the edge of evolution, especially laboratory results where evolutionary changes can be followed closely. The largest, most ambitious, controlled laboratory evolutionary study was begun more than a decade ago in the laboratory of Professor Richard Lenski at Michigan State University. Lenski wanted to follow evolution in real time. He started a project to watch the unfolding of cultures of the common gut bacterium *Escherichia coli*. *E. coli* is a favorite laboratory organism that has been studied by many scientists for more than a century. The bug is easy to grow and has a very short generation span of as little as twenty minutes under favorable conditions. Like those of *P. falciparum*, *H. sapiens*, and HIV, the entire genome of *E. coli* has been sequenced.

Unlike malaria and HIV, which both have to fend for themselves in the wild and fight tooth and claw with the human immune system, the *E. coli* in Lenski's lab were coddled. They had a stable environment, daily food, and no predators. But doesn't evolution need a change in the environment to spur it on? Shouldn't we expect little evolution of *E. coli* in the lab, where its environment is tightly controlled? No and no. One of the most important factors in an organism's environment is the presence of other organisms. Even in a controlled lab culture where bacteria are warm and well fed, the bug that reproduces fastest or outcompetes others will dominate the population. Like gravity, Darwinian evolution never stops.

But what does it yield? In the early 1990s Lenski and coworkers began to grow *E. coli* in flasks; the flasks reached their capacity of bacteria after about six or seven doublings. Every day he transferred a portion of the bugs to a fresh flask. By now over thirty thousand generations of *E. coli*, roughly the equivalent of a million years in the history of humans, have been born and died in Lenski's lab. In each flask the bacteria would grow to a population size of about five hundred million. Over the whole course of the experiment, perhaps ten trillion, 10^{13}, *E. coli* have been produced. Although ten trillion sounds like a lot (it's probably more than the number of primates on the line from chimp to human), it's virtually nothing compared to the number of malaria cells that have infested the earth. In the past fifty years there have been about a billion times as many of those as

E. coli in the Michigan lab, which makes the study less valuable than our data on malaria.

Nonetheless, the *E. coli* work has pointed in the same general direction. The lab bacteria performed much like the wild pathogens: A host of incoherent changes have slightly altered pre-existing systems. Nothing fundamentally new has been produced.[25] No new protein-protein interactions, no new molecular machines. As with thalassemia in humans, some large evolutionary advantages have been conferred by breaking things. Several populations of bacteria lost their ability to repair DNA. One of the most beneficial mutations, seen repeatedly in separate cultures, was the bacterium's loss of the ability to make a sugar called ribose, which is a component of RNA. Another was a change in a regulatory gene called *spoT*, which affected en masse how fifty-nine other genes work, either increasing or decreasing their activity. One likely explanation for the net good effect of this very blunt mutation is that it turned off the energetically costly genes that make the bacterial flagellum, saving the cell some energy. Breaking some genes and turning others off, however, won't make much of anything. After a while, beneficial changes from the experiment petered out.[26] The fact that malaria, with a billion fold more chances, gave a pattern very similar to the more modest studies on *E. coli* strongly suggests that that's all Darwinism can do.

THE PROTEIN EDGE

To put the difficulty of developing one or two protein-protein binding sites in perspective, Table 7.1 lists some approximate population sizes and likelihoods for some selected events. Figure 7.4 graphs results from the four dissimilar species: human, *E. coli*, HIV, and malaria. The number of cellular protein-protein binding sites developed by random mutation and natural selection for each (one for human—due to sickle cell hemoglobin—and zero for the other species) is plotted against the species' population size. The bottom axis of the graph extends from 10^0 to 10^{40}; since there likely have been fewer than 10^{40} organisms during the entire history of the earth, the bottom axis represents all of life.

TABLE 7.1
Some Approximate Population Sizes and Likelihoods for Selected Events

POPULATION SIZES—Total number of:

Primates in the line leading to modern humans in the past ten million years	10^{12}
Malaria cells in one sick person	10^{12}
Malaria cells worldwide in one year	10^{20}
Bacterial cells in the history of life on earth	10^{40}

LIKELIHOOD OF AN EVENT BY RANDOM MUTATION—Average number of:

Malaria cells needed to generate atovaquone resistance	10^{12}
Malaria cells needed to generate chloroquine resistance	10^{20}
Estimated organisms needed to generate one protein-binding site	10^{20}
Estimated organisms needed to generate two protein-binding sites	10^{40}

PROTEIN-BINDING SITES GENERATED BY RANDOM MUTATION:

Humans (10^8 organisms)	1
E. coli (10^{13} organisms)	0
HIV (10^{20} viruses)	0
Malaria (10^{20} organisms)	0

PROTEIN-BINDING SITES FOUND IN A TYPICAL CELL: 10,000

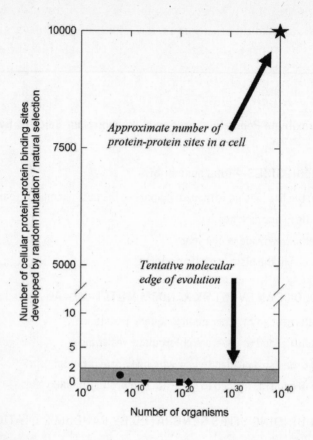

FIGURE 7.4

Graph of the number of protein-binding sites produced by random mutation and natural selection versus the population size for human (circle), *E. coli* (triangle), HIV (square), and malaria (diamond). (The values used for the population sizes are, respectively: 10^8, 10^{13}, 10^{20}, and 5×10^{21} — 10^8 is the approximate number of humans needed to produce a sickle hemoglobin mutation; 10^{13} is the total number of *E. coli* in the experiments of Richard Lenski; 10^{20} is the estimated number of HIV in the past several decades, worldwide; the value of 5×10^{21} is calculated from 10^{20} malaria cells each year for the past fifty years, the approximate time since chloroquine was introduced.) Humans developed one binding site (for sickle hemoglobin); the other species developed none. The top of the area shaded gray marks the molecular edge of evolution. Notice that the vertical axis is discontinuous. The star in the upper right marks the approximate number of different kinds of protein-protein binding sites in a typical cell. Extrapolating from the observational data shows random mutation accounts for very few of those sites.

Most people of course are familiar with the ordinary concept of an average; for example, the average of 0 and 40 is 20. In mathematics there is a concept called the geometric average. It is the number whose exponent is the average of the exponents of other numbers. For example, the geometric average of 10^0 and 10^{40} is 10^{20}. In a geometric sense, 10^{20} is midway to 10^{40}. So in a geometric sense, the observational data plotted in Figure 7.4 cover half of all life that has ever existed on earth, since the data extends past 10^{20} on the bottom axis. That allows us to be very confident in extrapolating from the data. The arrow in the upper-right-hand corner of the figure is directed toward a large point near 10,000, which represents the rough number of protein-protein binding sites in a typical cell. Somehow all those binding sites developed during the history of life. Straightforward extrapolation from the observational data plotted at the bottom of Figure 7.4 strongly indicates that random mutation accounts for very few of them.

Earlier in this chapter, using considerations from shape space, I conservatively estimated that the probability of developing a new protein-protein binding site by random mutation and natural selection would probably be on the order of a CCC—roughly the same difficulty as the development of chloroquine resistance in malaria, about one in 10^{20}. After looking at the results from work on *P. falciparum* and HIV, that estimate may now seem much too generous. CCCs do happen, if the population numbers supply them. In recent years chloroquine resistance has popped up a number of times independently. Yet HIV, despite having undergone more of at least some kinds of mutations than cells have experienced since the beginning of the world, has produced no new interactions between viral proteins. Nor has malaria developed new cellular protein-protein interactions. It seems the likelihood of developing a useful new protein-protein binding site is actually worse than a CCC.

Could the edge of evolution be as close as a single cellular protein-protein binding site, rather than two? After all, no new such interactions have been uncovered in malaria and HIV. Could it be that shape-space reasoning has significantly underestimated the difficulty of developing a single new binding site in the crowded, tightly regulated interior of a cell? That's possible, and we always have to keep in mind that these estimates are rough and will be revised as more information becomes available. Still, I think it's better to err on the side of caution, allow room for the odd exception like sickle hemoglobin, and draw the line at complexes of three kinds of proteins (that is, two binding sites), as I do in Figure 7.4.

So let's accept my earlier conservative estimation, and spell out some implications. The immediate, most important implication is that complexes with more than two different binding sites—ones that require three or more different kinds of proteins—are beyond the edge of evolution, past what is biologically reasonable to expect Darwinian evolution to have accomplished in all of life in all of the billion-year history of the world. The reasoning is straightforward. The odds of getting two independent things right are the multiple of the odds of getting each right by itself. So, other things being equal, the likelihood of developing two binding sites in a protein complex would be the square of the probability for getting one: a double CCC, 10^{20} times 10^{20}, which is 10^{40}. There have likely been fewer than 10^{40} cells in the world in the past four billion years, so the odds are against a single event of this variety in the history of life. It is biologically unreasonable.

With the criterion of two protein-protein binding sites, we can quickly see why stupendously complex structures such as the cilium, the flagellum, and the machinery that builds them are beyond Darwinian evolution. The flagellum has dozens of protein parts that specifically bind to each other; the cilium has hundreds. The IFT particle itself has sixteen proteins; even complex A, the smaller subset of IFT, has half a dozen protein parts, enormously beyond the reach of Darwinian processes. In fact, drawing the edge of evolution at complexes of three different kinds of cellular proteins means that the great majority of functional cellular features are across that line, not just the most intricate ones that command our attention such as the cilium and flagellum. Most proteins in the cell work as teams of a half dozen or more.

If the great majority of cellular protein-protein interactions are beyond the edge of evolution, it is reasonable to view the entire cell itself as a nonrandom, integrated whole—like a well-planned factory, as National Academy of Sciences president Bruce Alberts suggested. This conclusion isn't a "God of the gaps" argument. Nonrandomness isn't a rare property of just a handful of extra-complex features of the cell. Rather, it encompasses the cellular foundation of life as a whole.

8

OBJECTIONS TO THE EDGE

This chapter makes some important distinctions and addresses potential objections. It considers counterarguments to my attempt to define the edge of evolution—not philosophical ones, about the "other side" of that boundary, but technical and logical ones about the line itself. After that, at the end of the chapter, I cross the line.

In order to be as confident as possible about where to draw the line marking the edge of evolution, we have to take into account all the relevant data. Not all protein interactions can be lumped into the same category; we have to make careful distinctions and then check them against the relevant facts. One small point to note, for example, is that it's three or more *different* proteins binding specifically to each other that I assert is beyond Darwinian processes, not just three or more copies of the same protein. A number of proteins, like sickle hemoglobin, bind repeatedly to copies of themselves using the same binding site, like many copies of a single simple Lego part that can be stacked on each other. A "stack" of thousands of such proteins, all of a single type, is not beyond Darwinian possibility.

Another, more important point to note is that I'm considering just cellular proteins binding to other cellular proteins, not to foreign proteins. Foreign proteins injected into a cell by an invading virus or bacterium make up a different category.[1] The foreign proteins of pathogens almost always are intended to cripple a cell in any way possible. Since there are so many more ways to break a machine than to improve it, this is the kind of task at which Darwinism excels. Like throwing a wad of chewing gum into a finely tuned machine, it's relatively easy to clog a system—much easier than making the system in the first place.[2] Destructive protein-protein binding is much easier to achieve by chance.

More interesting than proteins that just gum up cellular defenses are those that allow a pathogen to take advantage of a host cell system. For example, cells have several intricate systems that control their shape, one of which is based on a protein called actin. Actin can form long fibers by assembling many copies of itself, another example of a Lego stack. However, the assembly of actin fibers is tightly regulated by other proteins in the cell, so that it only takes place at the proper time and place. Several kinds of bacteria and viruses subvert the Lego-stacking process for their own benefit by attaching to one of the control proteins, tricking it into thinking actin should be assembled on the pathogen. In effect, the invading pathogen hijacks a cell process, which helps it to spread.

While that's a fascinating and medically important process, the pathogen protein just triggers a pre-existing cellular mechanism. Like a tree limb that falls in the wind and hits the switch of a complex machine, turning it on, the pathogen protein does very little on its own. Darwinism can explain that aspect of the pathogen, but not the hijacked process it triggers. Like the development of antifreeze protein in Antarctic fish, such minimally coherent phenomena probably mark the far boundaries of what Darwinian processes can do in microbes.

AN EXCEPTION?

One apparently large exception to the difficulty of forming new cellular protein-protein interactions is sickle cell hemoglobin itself. Instead of needing several changes to make a new binding site, sickle hemoglobin needed just one. With just one change in its amino acid sequence, sickle hemoglobin developed a new binding site that allowed it to stick weakly to itself, and thus conferred resistance to malaria on Sickle Eve. Instead of needing a hundred billion billion people, the change required maybe just a hundred million. Why?

The reason is that the red blood cell is very unusual. Most other types of cells contain many different kinds of proteins, no one of which overwhelms the cell. Because its job is to carry as much oxygen as it can from the lungs to the tissues, by contrast, the red blood cell is stuffed with one protein, hemoglobin, the oxygen-transporting protein—hundreds of millions of copies of it. Although it contains a number of other kinds of proteins as well, about 90 percent of red blood cell protein is hemoglobin. The very high concentration of hemoglobin makes it a lot easier for interactions between hemoglobin molecules to have a noticeable effect. To understand why, let's go back to the swimming pool analogy and think about objects that fit each other, but poorly. On average they would perhaps spend about 1 percent of their time together. They are easily knocked apart, and then have to drift around for a long while until they accidentally came together again. Well, suppose in the pool we had not just one copy of those poorly fitting pieces, but millions. Now when the pieces stuck and then got knocked apart, they would have to drift for a lot less time to bump into another copy of their partner. Instead of searching for that one mate in the pool, the proteins would have millions to stick to. Because they would spend much less time searching for a copy of their partner, they'd spend a much larger fraction of time stuck together, even though their attachment was weak. If they were symmetrical like hemoglobin, with two identical sides, they could stick to a partner using each face, which could stick to another, and so on, until many of the copies congealed and gummed up the swimming pool. The bottom line is that if a protein is highly concentrated in a cell, as hemoglobin is in the red blood cell, a single, shape-changing mutation has a much better chance of making the protein stick to itself. Conversely, if hemoglobin were present at more typical protein levels, the sickle mutation wouldn't work.[3] At normal levels multiple amino acid changes would likely be needed to make hemoglobin stick to itself.

A more interesting example than sickle hemoglobin is the case of a protein abbreviated FKBP. The change of one particular amino acid (at position 36) in this protein causes the protein to bind to itself with moderate strength (about a hundred times more strongly than sickle hemoglobin). Using a technique called X-ray crystallography, which allows scientists to visualize almost every atom in a protein, this mutant proved very unusual:

> The interface between the two proteins is characterized by a remarkably extensive and complementary set of contacts suggestive of a bona fide protein-protein interaction rather than an artificial pairing. . . . Thus the interaction strikingly resembles natural high-affinity protein-protein interfaces. . . . This result suggests that the . . . substitution may . . . relieve an inherent steric hindrance to intermolecular association. . . . The discrete . . . change elicited by the F36M mutation is remarkable and, to our knowledge, unprecedented.[4]

In other words, it looked like the protein was *pre-engineered* to be complementary to itself, but was kept apart in the premutated version.[5] Switching amino acids in the mutation removed a blockage. In other words, the behavior of the protein FKBP was unlike anything encountered before. The close fit of the protein may mean that it is actually built to self-associate in nature under some circumstances that had previously escaped attention. It might be an example of Darwinian destruction (the scientists unwittingly undid a previous mutation). In any case, FKBP shows the need to be very cautious in interpreting a single experimental result. The subtle tasks of some proteins in the cell might require that they be poised to bind to each other. Mutating proteins as these scientists did could give us a false reading of the difficulty of the task facing evolution. To get a better understanding we should look beyond isolated results to the best general information on evolution we have.

ACCIDENTAL JIGSAW PUZZLES

In the last chapter I argued that design could be detected in the very fit of complex parts. But is that always true? Just the other day my six-year-old daughter knocked a vase off a shelf in our home, and it broke into several big chunks. The ragged breaks were complex. No other objects in our home or out of it matched them. Of course, the chunks fit perfectly together, yet they weren't individually designed. Here's another example. Suppose a rock fell into a puddle of water. During the night the water froze; a person who carefully removed the rock from the ice would see that the rock and the hole in the ice were exactly complementary to each other. They weren't designed to match each other by an intelligent agent, as automobile parts are, nor did we have to search through a huge shape-space library to find them.

The reason they fit so closely, of course, is that the process that made one part depended on the other part. The shape of the ice simply reflected the shape of the object that marked the boundary of the water; the water froze around the rock. As a vase breaks, the two sides of a crack are necessarily reflections of each other. A zig for one side is automatically a zag for the other. So in order to conclude that two closely matched parts were purposely intended to fit each other, not only do they have to be complex, but the process that made one has to have been independent of the process that made the other. One reason scientists initially hypothesized that antibodies "molded" themselves to the molecules they bound was that it seemed the easiest way to explain the match in shape—the shape of the antibody would be determined by the shape of what it bound. But when that simple explanation didn't pan out, further research revealed the elegant immune system, which independently and efficiently covers all of shape space.

With a couple of interesting exceptions,[6] protein-protein binding isn't the result of processes analogous to breaking a vase or water freezing around a complex shape. It arises either from searching a huge shape-space library, as the immune system does, or by some nonrandom mechanism.

GIVE ME JUST A LITTLE MORE TIME

Time has always figured prominently in Darwinian explanations. Although few changes can be noticed in our own age, Darwinists say, over vast stretches of geological time imperceptible modifications of life can add up to profound ones. It's no wonder that we don't see much coherent variation going on in the biology of our everyday world—evolutionary processes are so slow that a human lifetime is like a moment. The work on malaria and HIV upon which I base much of the argument for the edge of evolution has mostly been done in just the past fifty years. So how can it tell us anything reliable about what could happen over millions or even billions of years?

Time is actually not the chief factor in evolution—population numbers are. In calculating how quickly a beneficial mutation might appear, evolutionary biologists multiply the mutation rate by the population size. Since for many kinds of organisms the mutation rate is pretty similar, the waiting time for the appearance of helpful mutations depends mostly on numbers of organisms: The bigger the population or the faster the reproduction cycle, the more quickly a particular mutation will show up. The numbers of malaria cells and HIV in just the past fifty years have probably greatly surpassed the number of mammals that have lived on the earth in the past several hundred million years. So the evolutionary behavior of the pathogens in even such a short time as a half century gives us a clear indication of what can happen with larger organisms over enormous time spans. The fact that no new cellular protein-protein interactions were fashioned, that mutations were incoherent, that changes in only a few genes were able to help, and that those changes were only relatively (not absolutely) beneficial—all that gives us strong reason to expect the same for larger organisms over longer times.

Still, are the numbers we've examined enough? A hundred billion billion (10^{20}) malarial cells and HIV viruses is certainly a lot, but it's minuscule compared to the number of microorganisms that have lived on the earth since it first formed. Workers at the University of Georgia estimate that 10^{30} single-celled organisms are produced every year; over the billion-year-plus history of the earth, the total number of cells that have existed may be close to 10^{40}. Looked at another way, for each malarial cell in the past fifty years there have been about 10^{20} other microorganisms throughout history. Can we extrapolate from malaria and HIV to all of bacteria? To all of life?

Sure. We do of course have to be cautious and keep in mind that we are indeed extrapolating, but science routinely extrapolates from what we see happening now to what happened in the past. The same laws of physics that work here and now are used to estimate broadly how the universe developed over billions of years. So we can also use current biology to infer generally what happened over the course of life on earth. Since we see no new protein-protein interactions developing in 10^{20} cells, we can be reasonably confident that, at the least, no new cellular systems needing two new protein-protein interactions would develop in 10^{40} cells—in the entire history of life, as illustrated in Figure 7.4. The principle we use to make the extrapolation—that the odds against two independent events is the multiple of the odds against each event—is very well tested.

We can be even more confident of extrapolating over all of life, because in some ways HIV itself has mutated as much as all the cells that have ever existed on earth. The mutation rate of HIV (and other retroviruses) is at least ten thousand times greater than the mutation rate of cells. The much higher mutation rate of HIV gives it an evolutionary advantage over cells that increases dramatically if multiple changes are needed. For cells of higher organisms, each nucleotide of DNA has at most a one in a hundred million (10^8) chance of mutating.[7] The odds of getting any two particular nucleotides to change in a cell in the same generation is that number squared, or one in 10^{16}. Any good bookie could do the math to see that it would take about 10^{40} cells to generate all possible six-nucleotide mutations.[8] On the other hand, when HIV replicates, each of its nucleotides has a one in ten thousand (10^4) chance of mutating. Two particular nucleotides changing at the same time in the virus would have odds of that number squared, one in 10^8, and so on. So to generate all possible six-nucleotide mutations in HIV would require only 10^{20} viruses, which have in fact appeared on earth in recent decades. In other words, while we have studied it, HIV has run the gamut of all the possible substitution mutations, a gamut that would require billions of years for cells to experience. Yet all those mutations have changed the virus very little. Our experience with HIV gives good reason to think that Darwinism doesn't do much—even with billions of years and all the cells in the world at its disposal.

Incidentally, the results with HIV also shed light on the topic of the origin of life on earth. It has been speculated that life started out modestly, as viral-like strings of RNA, and then increased in complexity to yield cells. The extremely modest changes in HIV throw cold water on that idea. In 10^{20} copies, HIV developed nothing significantly new or complex. Extrapolating from what we know, such ambitious Darwinian early-earth scenarios appear to be ruled out.

E PLURIBUS UNUM

In trying to determine where lies the edge of evolution, I've relied heavily on one organism, the malarial parasite, with support from two other microbes, HIV and to a much lesser extent E. coli. Yet, even though malaria does attain enormous population sizes, still it's only one kind of organism. There are millions of species of animals, and many more species of plants and microbes. Is it possible that some other organism might have a greater evolutionary potential than malaria or HIV or E. coli? Could it be that, unluckily, the best-studied examples just happened to be evolutionary laggards? That some bacterium or plant hidden away in an unexplored forest or ocean could run Darwinian rings around the million-murdering death?

Yes, in a logical sense it is possible. One can never completely rule out the unknown. Bare possibility, however, is a poor basis for forming a judgment about nature. A rational person doesn't give credence to a claim based on bare possibility—a rational person demands positive reasons to believe something. Until an organism is found that is demonstrated to be much more adept than the malarial parasite at building coherent molecular machinery by random mutation and natural selection, there is no positive reason to believe it can be done. And the best evidence we have from malaria and HIV argues it is biologically unreasonable to think so.

What's more, there are compelling reasons to suppose that the results we have in hand for malaria and HIV are broadly representa-tive of what is possible for all organisms. In the past fifty years biology has unexpectedly shown that to a remarkable degree all of life uses very similar cellular machinery: With a few minor excep-tions the genetic code is the same for all the millions of species on earth; proteins are made of the same kinds of amino acids; nucleic acids are made of the same kind of nucleotides; and many, many other basic similarities. A biochemistry textbook typically observes, "Although living organisms . . . are enormously diverse in their macroscopic properties, there is a remarkable similarity in their bio-chemistry that provides a unifying theme with which to study them."[9]

The physical forces between proteins do not vary from organism to organism, nor does protein shape space depend on species. Since the criterion we are using to determine the edge of evolution is the development of specific protein-protein interactions, which is one of the most fundamental features of life, in that regard malaria is no different from any other organism.

Another possible objection is that malaria and HIV were just try-ing to get rid of poisons—to counter antibiotics—any way they could. Since the problem they were trying to solve is so narrow, it's not surprising (one might say) that changes were concentrated in a few proteins, and that nothing at all complex was produced. Yet that objection would run up against a contradiction. It is widely thought that when it first appeared, atmospheric oxygen itself was poison-ous to cells. But it is also widely thought in Darwinian circles that random mutation and natural selection allowed cells not only to tolerate the poison, but to construct enormously complex cellular mechanisms to take *advantage* of oxygen. Richard Dawkins opined that arms races build complex coherent machinery—where is the complex new machinery to deal with chloroquine? If Darwinism could spin gold out of once-deadly oxygen, why can natural selec-tion do nothing with modern antibiotics? The obvious answer is that the premise is wrong: Random mutation did not build either the complex cellular machinery of respiration or any other. Left to its own devices, mutation and selection produce the disjointed, lim-ited responses we see for the case of modern antibiotics.

ONE AT A TIME

The conclusion from Chapter 7—that the development of two new intracellular protein-protein binding sites at the same time is beyond Darwinian reach—leaves open, at least as a formal possibility, that some multiprotein structures (at least ones that aren't irreducibly complex, in the sense defined in *Darwin's Black Box*) might be built by adding one protein at a time, each of which is an improvement. But there are strong grounds to consider even that biologically unreasonable. First, the formation of even one helpful intracellular protein-protein binding site may be unattainable by random mutation. The work with malaria and HIV, which showed the development of no such features, puts a floor under the difficulty of the problem, but doesn't set a ceiling. Maybe my conservative estimate of the problem of getting even a single useful binding site is much too low. What we know from the best evolutionary data available is compatible with not even a single kind of specific, beneficial, cellular protein-protein interaction evolving in a Darwinian fashion in the history of life.

A second reason to doubt a one-protein-at-a-time scenario is the demanding criterion of coherence. The longer an evolutionary pathway, the much more likely that incoherent, momentarily-helpful-but-dead-end mutations will sidetrack things. The pathway to just one binding site is long, so the pathway to a second one is even longer. That means many more opportunities to take a wrong turn and get stuck on some tiny hill in a rugged landscape. As noted earlier, Allen Orr showed that on average just one or two steps would land an organism at a local evolutionary optimum, unable to progress further. Although Orr was discussing nucleotide sequences, it is reasonable to think the same consideration operates at other biological levels, so that at best one or two new protein-binding sites would present a local evolutionary peak, resistant to further change.

A third reason for doubt is the overlooked problem of restricted choice. That is, not only do new protein interactions have to develop, there has to be some protein available that would actually do some good. Malaria makes about fifty-three hundred kinds of proteins. Of those only a very few help in its fight against antibiotics, and just two are effective against chloroquine. If those two proteins weren't available or weren't helpful, then, much to the joy of humanity, the malarial parasite might have no effective evolutionary response to chloroquine. Similarly, in its frantic mutating, HIV has almost certainly altered its proteins at one point or another in the past few decades enough to cover all of shape space. So new surfaces on HIV proteins would have been made that could bind to any other viral protein in every orientation. Yet of all the many molecules its mutated proteins must have bound, none seem to have helped it; no new protein-protein interactions have been reported. Apparently the choice of proteins for binding is restricted only to unhelpful ones.

Restricted choice is a problem not only in fighting antibiotics, but also in fighting the environment and other organisms. Although malaria has only had a few decades to deal with manmade antibiotics, it has pretty much trashed them all, because only one or a few point mutations were needed. Yet it's had about ten thousand years to deal with the sickle hemoglobin mutation and has been unable to get around it. The same with other human genetic responses to malaria—thalassemia, hemoglobin E, and so on. It may be that there simply is no effective mutational response that is available to malaria. The same with its vulnerability to chilly temperatures. Even though Antarctic fish cobbled together an antifreeze system by random mutation to survive in icy waters, with many more chances P. falciparum hasn't learned to even knit itself a figurative sweater. As with sickle hemoglobin, it seems likely that there simply is no available evolutionary response. Nothing helps.

When you are building a fine-tuned, multicomponent cellular structure, the problem gets exponentially more severe at each step, as many specialized components are required. The bottom line is, it's reasonable to think that building multiprotein complexes one protein at a time is also well beyond the edge of evolution.

WHAT LIES BEYOND THE EDGE?

Although Darwin's is one theory of how unintelligent forces may mimic intent, it isn't the only one. So if random mutation and natural selection can't do the trick, maybe some other unintelligent process can. Although Darwin's theory is far and away most biologists' favored account for the appearance of design in life, a minority of biologists think it's woefully inadequate and prefer other unintelligent explanations.

One of the more popular minority views, called "complexity theory" or "self-organization," has been championed for decades by Stuart Kauffman, currently of the University of Calgary. The use of the term "self-organizing" can be a bit confusing because *all* of biology is profoundly self-organizing, as we saw with the example of IFT. But that's not what's meant here. Self-organization theorists use the term in a more general way. For example, one nasty example of self-organization from our everyday world is a hurricane—when conditions are right, the ocean, atmosphere, and heat combine to forge a highly organized storm that can persist for weeks. But most of the physical details of the system aren't critically precise. It's also completely unclear how the concept would apply to evolution. While it's certainly plausible that in some instances biological systems can self-organize in Kauffman's sense, there's no reason to think that self-organization explains how complex genetic systems arose.[10] Here's an illustration from everyday life. Some very simple rush hour traffic patterns are self-organizing, but self-organization does not explain where very complex carburetors, steering wheels, and all the other physical parts of a car came from, let alone how "cars could be manufactured by merely tumbling their parts onto the factory floor." In the same way intraflagellar transport might be self-organizing in the sense that it self-assembles, but self-organization doesn't explain how the structures that IFT depends on arose.

A second rival to Darwin has been dubbed "natural genetic engineering" by its most prominent proponent, University of Chicago biologist James Shapiro. The gist of the idea is that cells contain the same tools that human genetic engineers use to manipulate genes, to clone, and generally to tinker with life. In fact, in most cases that's where the human engineers got the tools—from cells. Cells have proteins that can cut pieces out of DNA, move them to different places in the cell, repeatedly duplicate genes, and so on. What's more, the cell itself "knows" where critical regions in the DNA are: where genes start and stop, which regions are inactive and which are active, and so on. The cell "knows" this because it contains proteins that sense all those features. Since cells contain sophisticated tools, the argument suggests, evolution doesn't have to proceed in a Darwinian manner by tiny random changes. It can progress in big steps, just as human genetic engineers take big steps when manipulating cells.

In many ways Shapiro has a higher, more respectful view of the genome than do Darwinists. Over the years, some Darwinists have derided portions of DNA where sequences are repeated many times as "junk." Shapiro disagrees:

> Despite its abundance, the repetitive component of the genome is often called "junk," "selfish," or "parasitic" DNA. . . . We feel it is timely to present an alternative "functionalist" point of view. The discovery of repetitive DNA presents a conceptual problem for traditional gene-based notions of hereditary information. . . . We argue here that a more fruitful interpretation of sequence data may result from thinking about genomes as information storage systems with parallels to electronic information storage systems. From this informatics perspective, repetitive DNA is an essential component of genomes; it is required for formatting coding information so that it can be accurately expressed and for formatting DNA molecules for transmission to new generations of cells.[11]

Shapiro thinks the genome is much more sophisticated than we had supposed; it's like a computer that contains not only specific programs, but an entire operating system. Shapiro's thinking makes random (although not "Darwinian") evolution more plausible, because the randomness includes steps that are more likely to be helpful.[12]

Unfortunately, in my view, natural genetic engineering proponents mistake cause for effect. Although big changes in repetitive DNA sequences certainly may affect gene expression[13] and animal shapes[14] (just as point mutations in proteins and more "traditional" Darwinian processes may do),[15] natural genetic engineering does not explain where the engineering tools came from, or how they can be employed coherently, or how formatting came about, or how it might change coherently, or a host of other pressing questions.

It's one thing to say that both Windows and Apple operating systems require formatting, and that they both have programs for copying, editing, and deleting computer code. It's quite another to say either that the codes arose by unintelligent processes, or even that Apple formatting could be switched to Windows formatting by a series of beneficial, random changes. Big changes in *Moby Dick*—duplicating chapters, rearranging paragraphs and sections—won't convert it into a new story any more than will small changes, such as spelling changes or duplicating or deleting single words. Monkeys typing on computers equipped with even the most advanced word-processing features still can't generate coherent changes to a text—only intelligence can. Shapiro makes a strong case that the genome is much more sophisticated than had been thought, and that changes in repetitive DNA can affect an organism. But if anything he is pointing the way to a possible mechanism for the unveiling of a *designed* process of common descent. Something must control this process; it cannot be random.

In fact, old-fashioned Darwinism demonstrably has more going for it than rival unintelligent theories. Self-organization and self-engineering played no visible role in the evolution of malaria and HIV over the past fifty years. Whatever the many shortcomings of Darwin's theory, small random mutations and natural selection do seem to account perfectly well for the resistance of malaria to chloroquine and of HIV to various drugs. On the other hand, in a hundred billion billion chances, there have been no apparent occasions where unintelligent-but-non-Darwinian processes have helped much—no sudden changes to new cellular states, no massive rearrangements by the genome reengineering itself. The only hint of non-Darwinian events in the best studies of evolution is found in Richard Lenski's work with *E. coli*.[16] He reported that "insertion sequences" (DNA sequences resembling viruses that can hop around a genome) do cause many mutations, which often break and disrupt genes. Yet none of them make new cellular structures. The fact that natural genetic engineering processes are indeed quite active, as Lenski and others have shown, yet malaria and HIV have made no good use of them in 10^{20} tries, strongly suggests they have very limited utility.

Indeed, the work on malaria and AIDS demonstrates that *all possible* unintelligent processes in the cell—both ones we've discovered so far and ones we haven't—at best have extremely limited benefit, since no such process was able to do much of anything. It's critical to notice that *no artificial limitations* were placed on the kinds of mutations or processes the microorganisms could undergo in nature. *Nothing*— neither point mutation, deletion, insertion, gene duplication, transposition, genome duplication, self-organization, self-engineering, nor any other process as yet undiscovered—was of much use. Darwinism helped the parasites a little bit, so it takes the prize for the best of the unintelligent mechanisms. But any other putative non-Darwinian, unintelligent processes were undetectable. It's reasonable to conclude, then, that all other unintelligent processes are even less effective than Darwinism.

DARWIN MEETS MICHELSON-MORLEY

P. falciparum, HIV, and *E. coli* are all very, very different from each other. They range from the simple to the complex, have very different life cycles, and represent three different fundamental domains of life: eukaryote, virus, and prokaryote. Yet they all tell the same tale of Darwinian evolution. Single simple changes to old cellular machinery that can help in dire circumstances are easy to come by. This is where Darwin rules, in the land of antibiotic resistance and single tiny steps. Burning a bridge that can stop an invading army or breaking a lock that can slow a burglar are easy and effective. But if just one or a few steps have to be jumped to gain a beneficial effect, as with chloroquine resistance, random mutation starts breathing hard. Skipping a few more steps appears to be beyond the edge of evolution.

There is much evidence from these studies that, in their incoherent flailing for short-term advantage, Darwinian processes can easily break molecular machinery. There is *no* evidence that Darwinian processes can take the multiple, coherent steps needed to build new molecular machinery, the kind of machinery that fills the cell.

Yet if it can do so little, why is random mutation / natural selection so highly regarded by biologists? Because the dominant theory requires it. There is ample precedent in the history of science for the overwhelming bulk of the scientific community strongly believing in imaginary entities postulated by a favored theory. For example, in the nineteenth century physicists knew that light behaved as a wave, but a wave in what? Ocean waves travel through water, sound waves through air; what medium do light waves travel through as they traverse space from the sun to the earth? The answer, announced with the utmost confidence by James Clerk Maxwell, the greatest physicist of the age, was the "*aether*" (that is, "ether").[17]

> Whatever difficulties we may have in forming a consistent idea of the constitution of the aether, *there can be no doubt* that the interplanetary and interstellar spaces are not empty, but are occupied by a material substance or body, which is *certainly* the largest, and probably the most uniform body of which we have any *knowledge*. (emphasis added)

In his article "Ether," published in the *Encyclopedia Brittanica* in the 1870s for all the world to read, the eminent Maxwell simply voiced the shared certainty of the entire physics community: Light was a wave, a wave needed a medium, the medium was called ether. In the encyclopedia article Maxwell not only proclaimed the existence of the ether, he precisely calculated its density and coefficient of rigidity! But in 1887 Albert Michelson and Edward Morley conducted a now-classic experiment to discern the presence of the ether, and found absolutely nothing. No trace of the "essential" substance. Whether the physicists' theories needed it or not, no ether could be detected.

Just as nineteenth-century physics presumed light to be carried by the ether, so modern Darwinian biology postulates random mutation and natural selection constructed the sophisticated, coherent machinery of the cell. Unfortunately, the inability to test the theory has hampered its critical appraisal and led to rampant speculation. Nonetheless, although we would certainly have wished otherwise, in just the past fifty years nature herself has ruthlessly conducted the biological equivalent of the Michelson-Morley experiment. Call it the M-H (malaria-HIV) experiment. With a billion times the firepower of the puny labs that humans run, the M-H experiment has scoured the planet looking for the ability of random mutation and natural selection to build coherent biological machinery and has found absolutely nothing.

Why no trace of the fabled blind watchmaker? The simplest explanation is that, like the ether, the blind watchmaker does not exist.

OVER THE EDGE

All unintelligent processes give very limited benefit. It's at this point in the book, then, that we must plunge across the boundary of Darwinian evolution to ponder what lies beyond. On this side of the edge of evolution lie random mutation and natural selection. On the other side—what?

First, it's certainly reasonable to suppose that natural selection plays a large role on both sides. After all, by itself natural selection is an innocuous concept that says only that the more fit organisms will tend to survive. Such a truism pretty much has to be operative in almost any biological setting. The big question, however, is not, "Who will survive, the more fit or the less fit?" The big question is, "How do organisms become more fit?" Or (now that we know much more about the molecular foundations of life) the question is, "Where did complex, coherent molecular machinery come from?" Even for that big question, the answer almost certainly will involve natural selection (at least after something has been supplied for natural selection to favor).

But just as certainly the answer will *not* involve random mutation at the center. From our best relevant data—parasitic diseases of humanity—we see that random mutation wreaks havoc on a genome. Even when it "helps," it breaks things much more easily than it makes things and acts incoherently rather than focusing on building integrated molecular systems. Random mutation does not account for the "mind-boggling" systems discovered in the cell.

So what does? If random mutation is inadequate, then (since common descent with modification strongly appears to be true) of course the answer must be *non*random mutation. That is, alterations to DNA over the course of the history of life on earth must have included many changes that we have no statistical right to expect, ones that were beneficial beyond the wildest reach of probability. Over and over again in the past several billion years, the DNA of living creatures changed in salutary ways that defied chance.

What caused DNA to change in nonrandom, helpful ways? One can envision several possibilities. The first is bare chance—earth was just spectacularly lucky. Although we have no right to expect all the many beneficial mutations that led to intelligent life here, they happened anyway, for no particular reason. Life on earth bought Powerball lottery ticket after lottery ticket, and all the tickets simply happened to be grand prize winners. The next possibility is that some unknown law or laws exist that made the cellular outcomes much more likely than we now have reason to suppose. If we eventually determine those laws, however, we'll see that the particular machinery of life we have discovered was in a sense written into the laws. A third possibility is that, although mutation is indeed random, at many critical historical junctures the environment somehow favored certain explicit mutations that channeled separate molecular parts together into coherent systems. In this view the credit for the elegant machinery of the cell should go not so much to Darwin's mechanism as to the outside world, the environment at large.

Each reader must make his own judgments about the adequacy of these possible explanations. I myself, however, find them all unpersuasive. Although much more could be said, briefly my reasons are these. The first possibility—sheer chance—is deeply unsatisfying when invoked on such a massive scale. Science—and human rationality in general—strives to explain features of the world with reasons. Although serendipity certainly plays its part in nature, advancing sheer chance as an explanation for profoundly functional features of life strikes me as akin to abandoning reason altogether. The second and third possibilities both seem inadequate on other grounds. They both seem in a sense to be merely sweeping the problem of the complexity of life under the rug. The second possibility replaces the astounding complexity of life with some unknown law that itself must be ultracomplex. The third possibility simply projects the functional complexity of life onto the environment. But, even in theory, neither the second nor third possibilities actually reduce complexity to simplicity, as Darwin's failed explanation once promised to do.

Instead, I conclude that another possibility is more likely: The elegant, coherent, functional systems upon which life depends are the result of deliberate intelligent design. Now, I am keenly aware that in the past few years many people in the country have come to regard the phrase "intelligent design" as fighting words, because to them, the word "design" is synonymous with "creationism," and thus opens the door to treating the Bible as some sort of scientific textbook (which would be silly). That is an unfortunate misimpression. The idea of intelligent design, although congenial to some religious views of the universe, is independent of them. For example, the possibility of intelligent design is quite compatible with common descent, which some religious people disdain. What's more, although some religious thinkers envision active, continuing intervention in nature, intelligent design is quite compatible with the view that the universe operates by unbroken natural law, with the design of life perhaps packed into its initial set-up. (In fact, possibilities two and three listed above—where nonrandomness was assigned either to complex laws or to the environment—can be viewed as particular examples of this. I think it makes for greater clarity of discussion, however, just to acknowledge explicitly in those cases that the laws or special conditions were purposely designed to produce life.)

In the remainder of the book, I'll plainly treat the other side of the edge of evolution as the domain of design. Readers who strongly disagree with design may take it simply as showing how the design argument is framed, or just as showing how little Darwinian processes explain and how much is not understood. Readers who are open to design explanations can see how well it fits with other aspects of nature that science has recently uncovered.

SURVIVOR

What is the rational justification for chalking up to design features of life that may be just barely over the edge of evolution, such as molecular systems that contain two different cellular protein-protein binding sites? After all, up until now I have shown simply that it was biologically unreasonable to think that Darwinian processes produced them. How do we proceed from the improbability of Darwinism to the likelihood of design?

Let's consider an analogy. Imagine that, like Tom Hanks in *Cast Away*, you wash up on a tropical island, the sole survivor of a plane crash. Choking and spitting out water, you pull yourself up off the sand and set off to explore the island, to look for food and shelter. After hours walking along the beach catching crabs, you turn and head for the interior mountains, hoping to find a cave to use as a base. Eventually you stumble across a sizeable crevice in the side of a mountain where you can at least take cover from storms. Over the next few weeks you range farther and farther on the large island, finding some coconut trees here and other edible plants there.

One day while exploring a distant stony beach you notice a half dozen football-sized rocks close together, forming a small crescent. Odd. But there are a lot of rocks around and they have to be in some pattern, so why not a crescent? About fifty yards away on the edge of the same beach you find another group of rocks, roughly the same size as the first, but this group has a couple dozen rocks and forms a complete circle, about four feet in diameter; no other rocks are close by. Very odd. Maybe a freak accident. Maybe a larger rock got hit by lightning, shattering it into pieces that landed in a circle, or possibly a swirling wave pushed rocks into a circle.

A week or two later while exploring the jungle, you spot a banana tree. Overjoyed at the prospect of a new food source you continue in the same direction, hoping to find a few others. During a ten-minute walk you find some more banana trees, a few scattered—then six of them, in two rows of three, each spaced about a dozen feet apart. Strange. Why should they grow like that? Were there just three original trees that happened to be growing in a row, and then perhaps a steady wind blew seeds perpendicular to the row? Or maybe there were two original trees, and the wind blew seeds to make two rows of three? But what sort of a lucky wind would it take to space the seeds so evenly?

A little farther into the jungle you find a grove of thirty-five mango trees in five neat rows of seven. About a quarter mile from the grove you discover a square of stone walls, with four straight sides ten feet long, each with three layers of stone neatly atop each other. Running now, you surmount a hill and for the first time spy the other side of the island. On the far beach, broken and weathered, are the remains of a small sailing vessel, a hundred years old by the looks of it. Its mast is snapped, planks are missing from the hull, and only shreds of the sail remain.

After rummaging through the ship, you walk back to your cave, and again pass the banana and mango groves, the square of rocks, and the circle and crescent. Now you see them differently. Did the wind blow seeds into neat rows of fruit trees, or did a shipwrecked sailor plant them? How about those piles of stones? Not just the big square, but the circle and the crescent, too? Once it's crystal clear that some things on the island—the ship and its contents—are the result of intelligent design, you have to reevaluate other features of the island. Now possible explanations include not only nature and luck, but mind and purpose, too.[18] Yet how do you decide if something is more likely accounted for by intelligence rather than the natural forces that also are at play on the island?

Here's one way. Design is *the purposeful arrangement of parts*.[19] Rational agents can coordinate pieces into a larger system (like the ship) to accomplish a purpose. Although sometimes the purpose of the system is obscure to an observer who stumbles upon it, so the design goes unrecognized,[20] usually the purpose can be discerned by examining the system. What's more, the arrangement of the pieces is frequently one that is quite unlikely to occur by chance. So if something on the island now looks as if it might have served some palpable objective, and if it seems quite unlikely to be the result of chance, you decide that the best explanation may be that it was purposely arranged that way.

With those considerations in mind, you now judge that the mango grove is very likely to have been purposely planted. The purpose would be to provide a supply of food, and the probability of the mango trees' growing in five neatly spaced rows of seven seems quite low. However, although suspicious-looking, the two rows of banana trees might just be a coincidence. Flukes do happen, even when an intelligent agent is around, so it's hard to tell for sure. The crevice in the mountain you are using as a base is not an uncommon natural occurrence—no reason to suspect design there. The square, three-layered stone pile is presumptively an incomplete or dismantled makeshift shelter; the four-foot circle of rocks is most probably some old campsite, rather than the aftermath of a lightning strike. But instead of a second campsite where some of the rocks were washed away, the small crescent of rocks might really be a fortuitous arrangement. After all, there are a lot of other rocks around, and some simple pattern or other might pop up just by chance.

You would make inferences based on your experience of the likelihood of some event happening by chance. You might be wrong in some cases, when your estimation is off. What's more, new evidence (such as discovering that the crescent is actually part of a large circle of rocks—a second campsite—the rest of which were covered by sand) could affect a conclusion, just as the new evidence of the discovery of the ship affected your judgments. Your level of confidence in design for different cases could range widely, from sneaking suspicion to utter certainty. As the estimated probability of serendipity decreases and the clarity of the purpose of the arrangement increases, your confidence in design would also increase. The stone crescent may be a fluke; the makeshift shelter almost certainly isn't. The wrecked ship itself, never.

As for a marooned fellow exploring an island, so, too, for biologists probing the hidden corners of life. In the past half century science has made enormous strides in understanding the molecular basis of life. In terms of the island illustration above, in the past few decades science has surmounted that final hill and spied stunning examples of design where it hadn't been expected, in the cell. For those who don't rule it out from the start, design is as evident in such sophisticated systems as the cilium as it is for the castaway in the wrecked ship. Once design has been established for such luminous cases, it then becomes a possible explanation for other, less overpowering examples. There will always be hard cases in the middle, but using the same principles as the stranded gent, we can go back and reappraise many features of life on earth. If a cellular feature has some discernible function, and if it seems to be beyond what is biologically reasonable to expect of chance, then with varying degrees of confidence we are justified in chalking it up to design.

BEYOND MOLECULES

Design dominates the molecular level of life. But what of higher levels of biological organization, beyond the cell? What of animal body shapes? Mammals versus fish? Individual species?

9

THE CATHEDRAL AND
THE SPANDRELS

HOW DEEP GOES DESIGN?

Up until now we have examined molecular structures and processes and have drawn a tentative line marking the molecular edge of Darwinian evolution. Most protein-protein interactions in the cell are not due to random mutation. Since cells are integrated units, it's reasonable to view cells in their entirety as designed. But keep in mind that accidents do happen, so there are Darwinian effects, of some degree, everywhere. For example, just as automobiles may accumulate dents or scratches over time, or have mufflers fall off, but nonetheless are coherent, designed systems, so, too, with cells. Some features of cells of course result from genetic dents or scratches or loss, but the cell as a whole, it seems, was designed.

Now it's time to look at higher levels of biological organization. There are several major classes of cells, which include the simpler prokaryotic cells of bacteria and the more complex eukaryotic cells of creatures ranging from yeasts to humans. Were just the simpler, prokaryotic cells designed? Could the more complex eukaryotic cells have evolved from them over time by unintelligent processes? In other words, given the simpler, designed cells in the distant past as a starting point, is it biologically reasonable to think that random mutation and natural selection could reach the more complex cells?

No. Eukaryotic cells contain a raft of complex functional systems that the simpler prokaryotes lack, systems that are enormously beyond Darwinian processes. For example, the cilium discussed in Chapter 5, which contains hundreds of protein parts, and IFT, the system that constructs the cilium from the ground up, both appear in eukaryotic cells, but not in prokaryotic cells. And the cilium isn't the only difference. As the evolutionary developmental biologists Marc Kirschner and John Gerhart exclaim in *The Plausibility of Life,* "enormous innovations attended the evolution of the first single-celled eukaryotes one and a half to two billion years ago."[1] The innovations include such fundamental features as sexual reproduction (meiosis and recombination), the organization of DNA into chromatin, and the provisioning of a cellular protein "skeleton." Of course, the two kinds of cells share a number of similar systems, such as the genetic code. Nonetheless, just as it's reasonable to view a motorcycle as a different sort of system from a bicycle, because eukaryotic cells contain multiple complex systems that prokaryotes do not, it's reasonable to view eukaryotes as integrated, designed systems in their own right,

So design extends beyond the simplest cells at least to more complex cells, which is the biological level of "kingdom." Does it go further? Although prokaryotes are single-celled organisms, not all eukaryotes are. Eukaryotes include not only single-celled organisms such as yeast and malaria, but also multicellular organisms: plants, and animals from jellyfish to insects to humans. So does design stop at the eukaryotic cell, or does it extend to multicellular organisms? More pointedly, given a generic, designed, eukaryotic cell in the distant past, is it biologically reasonable to think that over time the rest of life developed from it entirely by unintelligent processes? This chapter answers that question.

Before we begin, I should be clear that the arguments of this chapter will necessarily be more tentative and speculative than for previous chapters, which dealt with molecules and the cell. The reason is simply that, although rapid progress is being made, much less is known about what it takes to build an animal than about what it takes to build a protein machine. No experiments like those of Greg Winter exploring the shape space of proteins have been done to, say, thoroughly explore the shapes of animals. What's more, to be secure in our conclusions about life—even about large animals—we have to understand the relevant biology *at the molecular level*. The inflexible fact is that all of physical life is built of molecules, whose intricate interactions make possible such things as plants and animals. Like a computer, whose overall shape is visible to the naked eye but whose basic workings take place in microscopic circuits, animals live or die depending on the workings of invisible molecular machines. So to locate the edge of evolution, we have to understand the molecular differences between levels of life.

Even just twenty years ago such a project would have been impossible, since little was known then about the molecular basis of animal life. But especially in the past decade an avalanche of information about the embryonic development of higher organisms has exploded into view. The information in hand isn't yet enough to allow us to draw definitive, quantitative conclusions. Nevertheless, Darwinian defenders have already begun using the new work to speculate freely about how their theory might still be salvaged (at least for higher levels of biology, beyond the cell). An entire field of inquiry has arisen in the past decade, appropriating the spectacular findings of developmental biology for evolutionary theory. It is the Darwinists' latest line of defense. Yet, as we'll see, the new work offers further evidence of design, extending up past animal body plans and the major branches of life.

A MOLECULAR SWITCH

Although Charles Darwin was a perceptive man, the molecular basis of animal development was hidden from him, as it was from all scientists of his age. When Darwin mused about how a bear might turn into a whale, or a light-sensitive spot into a full-fledged eye, he did so unhindered by knowledge of what would be needed for such transformations to occur. For a century after Darwin died, only inklings of the process arose as biologists investigated life. Reports of misshapen animals with missing or extra limbs or organs titillated scientific curiosity, but the beginnings of genuine understanding awaited the discovery of the molecular foundations of life. Once the molecular structure of DNA was unveiled in the 1950s, some of the necessary conceptual foundation was laid. The fog was gradually lifting; now science understood somewhat more clearly how molecules went about performing the necessary tasks of life.

A huge breakthrough in understanding how proteins control DNA and life came with the work of François Jacob and Jacques Monod in the 1960s. It was known then that bacteria could digest different types of sugars, including the most common kind, called glucose, as well as another, much less common sugar, called lactose, which is found in milk. Intriguingly, when bacteria were grown in the presence of glucose, they couldn't use lactose. Only in the absence of glucose and the presence of lactose could they digest the milk sugar. When glucose was missing, the bacteria made proteins that could pull lactose into the cell and metabolize it, but when no lactose was around, the bacteria didn't make those proteins. This was a very clever trick that made great biological sense, since in normal conditions the bacterium would waste energy if it manufactured proteins that could metabolize only a rarely encountered sugar. The interesting question was, How did the bacteria "know" when to switch on the genes for making the proteins?

Jacob and Monod discovered a defective mutant bacterium that made lactose-using proteins all the time, even in the absence of lactose. It was lacking a control mechanism. The French scientists reasoned that the bacteria contained another, hidden protein, which they called a "repressor." They conjectured that the repressor would ordinarily bind to a specific sequence of DNA near the genes that generated the lactose-using proteins, switching them off. In the presence of lactose, the milk sugar would bind to the repressor itself, changing the protein's shape enough to make it fall off the DNA, switching back on the previously blocked genes. Jacob and Monod surmised that the mutant bacteria had a broken repressor.

Their model turned out to be exactly correct, earned them a Nobel Prize, and blazed the path for understanding how the genetic program contained in the DNA of all organisms is controlled. There are three critical lessons of the Jacob-Monod model, which we now

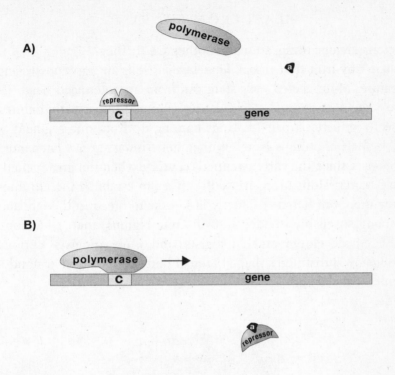

FIGURE 9.1
A simple genetic switch. (A) A repressor binds tightly to the control region (*c*)
of a gene, physically excluding the polymerase (which "transcribes" the gene)
from binding.(B) An activator (the small shape marked *a*) binds to the repressor, distorting its shape and causing it to fall off the gene, which allows the
polymerase to bind and begin transcription. (For simplicity, the role of glucose
and the CAP protein are not pictured.)

know apply not just to bacteria but to all of life: First, the genes for
many proteins in the cell aren't on all the time—they have to be
turned on or off at some point. Second, it is the job of some proteins
to control when the genes for other proteins are turned on and off.
The control proteins do little else in the cell other than to act as
molecular fingers to flip genetic switches. And third, there are
regions of DNA—usually close to the genes for the proteins that
they control—to which the control proteins bind. The physical
association of the control proteins to the DNA regions constitutes
the flipping of the switch.

BLAST FROM THE PAST

Bacteria are one thing, animals another. Or are they?

The tiny fruit fly *Drosophila melanogaster* is an unprepossessing creature. Multifaceted eyes stare out from an antennaed head, its body like a horizontal stack of tires chopped into clearly defined insect segments, a pair of wings coming up from one segment, a nubby pair of stumps from another. Yet *Drosophila* has enchanted biologists since the early twentieth century, when the great geneticist Thomas Hunt Morgan used the flies to establish the chromosome theory of heredity. The fly is so easy to breed in the lab, and its body so visibly divided into discrete regions, that it has long attracted developmental biologists and embryologists curious about how a distinctly shaped animal body is built from a nondescript fertilized egg.

FIGURE 9.2

The fruit fly *Drosophila melanogaster*. (Modified from Plate V in The University of Texas Publication No. 4313: April 1, 1943, Studies in the Genetics of *Drosophila* III. The *Drosophilidae* of the Southwest, Directed by J. T. Patterson, Professor of Zoology, The University of Texas. Courtesy of FlyBase.net.)

By crossbreeding a very large number of fruit flies in the 1970s, the Cal Tech geneticist Edward Lewis showed that the DNA in one region of one of *Drosophila*'s chromosomes contained a number of genes that appeared to regulate the development of different regions of the body of the fly. Curiously, the genes appeared to be arranged on the chromosome in the same order as the segments of the fly that they helped control, ranging from genes controlling development of head parts at the leftmost, genes for the thorax in the middle, and

genes for the abdomen at the right. Mutations in these genes sometimes had bizarre effects, including the formation of flies with four
wings instead of two, or flies that had legs emerging from their
heads where antennae should have been. Such monstrous alterations, which caused different sections of the animal's body to be
mixed up, were dubbed "homeotic" mutations. The important biological point was that one or a few mutations could cause big mix-
ups in the body plan of the animal.

In the 1970s and 1980s the German Christiane Nüsslein-Volhard
and American Eric Wieschaus used chemicals to mutate flies, and
in a heroic effort studied tens of thousands of different mutant
flies. From these they discovered there were more than a hundred
genes essential to fly development. Mutations in these genes didn't
cause a fly to just keel over and die. Rather, they caused big mix-ups
in the basic shape of its body. In the case of some mutations, whole
organs such as eyes were missing. With other mutations, the poor
fly embryo had only half as many body segments as usual. Clearly
these genes were not ones that coded for ordinary proteins like
hemoglobin. Apparently, the genes controlled long chains of events
leading to the building of large, discrete chunks of the fly's body.

But what exactly were those genes? By the mid-1980s biologists
could routinely determine the nucleotide sequence of fragments of
DNA. If the piece of DNA was part of a gene coding for a protein (as
opposed to "junk" DNA), the amino acid sequence of the protein
could be deduced directly from it and compared to the sequences of
other proteins. One of the first homeotic fruit fly genes sequenced,
in fact, coded for a protein that resembled the bacterial repressor
protein that Jacob and Monod studied in the 1960s[2]

That was a strong clue that, like the bacterial gene, the fly gene
also acted as part of a switch to turn other genes on or off. Surveys of
other organisms, ranging from worms to people, unveiled a whole
new class of such proteins, all containing a region of about sixty
amino acids similar to the repressor protein and very similar to one
another. The segment of the genes that coded for the sixty-amino-
acid region of a homeotic protein is called the "homeobox." The
proteins are dubbed Hox proteins.

In subsequent years homeotic proteins, and other classes of control proteins, have proven to be master regulators of developmental programs in animals. Although they resemble the repressor protein that Jacob and Monod discovered decades earlier, in that they bind near a gene to turn it on or off, the regulatory systems of animals are much, much more complex than bacterial systems. The bacterial lactose system was turned on or off by a single protein. In animals, a master switch sets in train a whole cascade of lesser switches, where the initial regulatory protein turns on the genes for other regulatory proteins, which turn on other regulatory proteins, and so on. Eventually, after a pyramid of control switches, a regulatory protein activates a gene that actually does some of the construction work to build an animal's body. But there's another complication. A gene in an animal cell might be regulated not by just one or a few proteins, as bacterial genes are, but by more than ten. What's more, there may be dozens of sites near the gene at which the regulatory proteins might bind, with multiple separate sites for some regulatory proteins.

TO BUILD A FLY

Why such enormous complexity, far beyond that of bacterial cells? The reason is that animal bodies contain many different kinds of cells that have to be positioned in definite relationships to other cells, in order to be formed into organs, and to connect to other parts of the body. Cal Tech biologist Eric Davidson emphasizes what the task of building an animal demands:

> The most cursory consideration of the developmental process produces the realization that the program must have remarkable capacities, for development imposes extreme regulatory demands . . . Metaphors often have undesirable lives of their own, but a useful one here is to consider the regulatory demands of building a large and complex edifice, the way this is done by modern construction firms. All of the structural characters of the edifice, from its overall form to minute aspects that determine its local functionalities such as placement of wiring and windows, must be specified in the architect's blueprints. The blueprints determine the activities of the construction crews from beginning to end.[3]

In other words, *the molecular developmental program to build an animal must consist of many discrete steps and be profoundly coherent.* As we've seen throughout this book, random mutation cannot take multiple coherent molecular steps. Therefore, like a castaway re-evaluating structures on an island in light of the knowledge that some things there were designed, we should *already* suspect that to some extent animal forms were designed. But to what degree?

For a flavor of the careful planning that goes into building even a relatively simple animal, let's look briefly and sketchily at some of what's been learned from studies of *Drosophila* development in recent years.[4] The mother fly starts the process off by depositing in the egg, at the end that will become the head, a concentration of the instructions[5] to make one kind of protein, called "bicoid," and, at the end that will become the tail, a second kind of protein, called "nanos." The bottom of the embryo is marked by the mother fly in a somewhat different way. The genes coding for proteins that specify the sides of the egg (front, back, top, bottom) are called "egg-polarity" genes. Critically, the proteins (or other proteins they affect) can stray in the egg, drifting away from their source; as they do they become more diffuse. As the egg initially divides into many cells, the high concentration of signal protein at one end of the fly turns on one set of control genes, the middle concentration turns on a different set of control genes in the middle portion of the embryo, and the lowest concentration activates a third set.

Once the front, back, top, and bottom are marked (caution—it's critical to keep in mind that the signal genes don't actually form the structures found in those regions of the developing fly; they simply mark the location of a cell, like a surveying crew mapping out land for a construction project), positions are further refined with other control proteins. Several groups of proteins controlled by "segmentation genes" subdivide the embryo further. One group of about six so-called "gap genes" is switched on, marking chunks of segments; if one of these control proteins is defective, several neighboring segments of the embryo will be missing. Oddly, another group of eight genes, called "pair-rule genes," affect alternate segments. If one of these is broken, a fly embryo will have only half its normal complement of segments. Finally, a group of ten "segment-polarity" genes helps differentiate each segment. Although in a normal fly the front of each segment looks a bit different from the back, in some seg-ment-polarity mutants the two ends look the same.

The details can be mind-numbing, but the shape of the process is important: from egg-polarity genes to gap genes to pair-rule genes to segment-polarity genes, and we still aren't ready to build the fly. The lifespan of all of the proteins coded by these control genes is brief, but they turn on genes for the more permanent Hox proteins, and thus permanently mark the position of cells in the developing animal embryo.

Similar processes subsequently lay out compartments at finer and finer levels of the fly. For example, as a wing is built, the front, back, top, and bottom are marked by control genes, sometimes the same control genes that earlier marked various regions of the entire embryo. But now, working in a defined region of the developing ani-mal, they mark the divisions and edges of the subcompartment. Remember, individual control genes don't by themselves embody the instructions to build a wing—they just mark areas of the fly, and signal other genes to turn on or off.

This short description leaves out many, many known details of the developmental process, including other means of cell-cell com-munication and the mechanics of how a signal is received and inter-preted. But it at least gives a taste of how the body plan of a simple organism is set in motion.

FLY BOY

The discovery of master regulatory molecules such as Hox proteins that controlled whole body sections of *Drosophila* was surprise enough. But researchers were absolutely astounded when the pro-teins were compared with those of distantly related organisms such as, say, people. Every Hox gene seen in the fruit fly has a very similar counterpart in humans! The similarities went well beyond the amino acid sequences of the proteins. The human counterparts even con-trolled the development of analogous sections of the developing human embryo. That is, the human counterpart to the fruit fly gene that controls the growth of insect head parts directs construction of regions near mammals' heads (the genes of all mammals are similar to those of humans). The tail end of humans is built under the direction of the mammalian counterpart of the master fly regulatory gene that directs the arrangement of the insect's hindquarters. Even more strangely, as with the fly, the genes in mammals were still lined up with body segments, with the leftmost gene coding for head regions, middle genes for middle-body regions, and rightmost gene for tail-end sections.

It seemed that life was imitating art with a vengeance. In the 1986 remake of the classic horror movie *The Fly,* a scientist accidentally mixes his DNA with that of the insect, and over the length of the film slowly and dramatically turns into a fly. The discovery that humans and *Drosophila* share the same master regulatory genes conjured visions of a person under a full moon sprouting wings and antennae.

The spooky dreams took a step toward reality in the 1990s when the Swiss biologist Walter Gehring isolated the corresponding gene from the fruit fly that was known to affect the development of eyes in vertebrates. Using clever lab techniques, he inserted a mouse gene into different spots in a developing fly embryo, and eyes grew in those spots![6] There were eyes on antennae, eyes on legs. To everyone's relief, the eyes at least weren't mouse eyes—they were the regular compound eyes of an insect. This re-emphasized that the master regulatory genes are simply switches, turning on the cellular hardware that does the actual construction of the organ. Just as the same kind of light switch can be used to turn on either an incandescent light or a fluorescent light, whose structures and mechanisms are considerably different, the eye gene just switches on the construction program in an animal that builds an eye.

Nonetheless, the result vividly brought home two points. First, animals as disparate as mice, flies, and worms rely to a very surprising degree on similar developmental programs that use similar components. (As a great practical benefit, this makes it possible to study the development of lower animals and use those results to infer biological facts about humans, where such experiments would be morally problematic.) Second, genetic programs to build organs such as eyes, limbs, and body segments seem to occur in discrete modules. After all, it took just one gene on some fly's leg to trigger the building of an eye where it shouldn't be. The rest of the genetic program clearly was already there, waiting to be activated.

This finding has two implications for Darwinism. First, it offers yet more confirmation of common descent. If mammals and flies use the same switching genes, it is reasonable to think that they inherited them from the same ancestor or ancestors. Second, it is possible for single mutations to have very large effects on animal bodies, rearranging whole regions in one fell swoop. So if under some odd circumstance it would be beneficial for a fruit fly to have an extra pair of eyes on its antennae, the eyes wouldn't have to be built from scratch, one tiny mutation at a time, first changing one protein in the antenna to something like rhodopsin, then changing another protein to start to form a lens, and so on. Maybe instead the gene for the master eye regulatory protein might by accident simply be switched on in an antenna cell, allowing a mutant animal to form extra eyes in a single generation. Or, less dramatically, perhaps extra legs or wings could be grown on body segments where they normally are missing, or suppressed where they usually occur.

With the discovery of master genetic regulatory programs for animal body modules, it seemed a viable path had opened up around Darwin's tedious insistence that evolution must always be gradual. Instead of changing letter by letter, now monkeys could rearrange whole chapters at a time. Now random mutation and natural selection could work by leaps and bounds.

MODULATING DARWIN

The recent exciting advances in understanding the genetic basis of animal embryology have helped spark a new field of inquiry dubbed "evolutionary developmental biology" or "evo-devo," for short. Evo-devo looks both at how animals are built in each generation and at how they might have evolved over millennia. Proponents of evo-devo typically whistle gingerly past questions of how basic cellular machinery may have come about by unintelligent processes at the start. But, given a generic eukaryotic cell that has been endowed with what's been styled a "tool kit" of regulatory genes, they imagine they can scout a path for mutation and selection to go from such humble creatures as flatworms, past insects and arachnids, up through fish, all the way to cats.

The dominant theme of the new thinking is "modularity."[7] As proponents admit, the concept can be pretty fuzzy. Roughly, a module is a more-or-less self-contained biological feature that can be plugged into a variety of contexts without losing its distinctive properties. A biological module can range from something very small (such as a fragment of a protein), to an entire protein chain (such as one of the subunits of hemoglobin), to a set of genes (such as Hox genes), to a cell, to an organ (such as the eyes or limbs of *Drosophila*). Some thinkers even apply the concept to mind, art, and culture.[8] In the next few sections I'll concentrate on the sustained discussions of modularity that I think are the most evolutionarily relevant. The bottom line is that, while great progress has been made toward understanding how animals are made, and has revealed unexpected, stunning complexity, no progress at all has been made in understanding how that complexity could evolve by unintelligent processes.

MANY SWITCHES, NO EXPLANATIONS

In *Endless Forms Most Beautiful,* University of Wisconsin biologist and leading evo-devo researcher Sean Carroll delivers a vivid, enthusiastic, firsthand account of the pioneering work of his own lab and others on fruit flies and butterflies. After discussing the discovery and action of Hox proteins and other regulatory proteins, Professor Carroll concentrates on what he believes to be the key to understanding animal evolution, which is the short DNA regions to which the regulatory proteins bind, which he calls "switches." Switches can be considered modules that can be placed next to any gene. Because each different kind of regulatory protein has a unique, relatively short (about six to ten nucleotides) "signature" sequence of DNA to which it binds near a gene that it helps to turn on, Carroll proposes that genes can be turned on and off over evolutionary time just by random mutations in the DNA region next to the gene.

The way it might work in evolution is something like the following. Suppose it would be beneficial for a developing structure (say an incipient wing or claw) in some evolutionarily promising creature to have a particular one of the ten thousand or so proteins in its genome turned on or off, or even just turned on more or less strongly (perhaps that would make a protein that on balance would strengthen the appendage). To do so it wouldn't have to evolve a brand-new protein just for the novel appendage. Instead, the region of DNA near the gene would just have to mutate a few nucleotides to form a switch region that could bind the correct one of the hundreds of regulatory proteins the animal's genome codes for. When the correct regulatory protein bound, perhaps the gene would be turned on or off—not in the whole animal, which might be damaging, but just in the subset of cells that form the appendage.

That sounds easy enough, and Carroll generally stops the story there. But, since one change surely would not give a different new structure, let us continue thinking along the same lines. Suppose a second protein would help push the process along. Well, then, like the first, just the right one of ten thousand genes would again have to develop just the right one of hundreds of possible switch regions. But what if, in the meantime, it would "help" to break a gene, as in thalassemia, which would occur hundreds of times more frequently than specific point mutations? Or what if a momentarily helpful but disconnected change popped up, as in hereditary persistence of fetal hemoglobin? Or what if a coherent change would require passing through a detrimental mutation, as with chloroquine resistance in malaria? What is the likelihood of those looming brick walls?

The scenarios in Carroll's book seem persuasive because they focus on a single switch or protein. Like considering just one short sentence ("Call me Ishmael") of a much longer literary work, zeroing in on just one aspect of a difficult evolutionary problem reduces what in reality is a very rugged landscape to one that apparently consists of a single gentle evolutionary hill. From a broader perspective, however, the evo-devo process looks as if it has as much potential for incoherence—with successive evolutionary steps jumbled and disconnected from each other—as traditional Darwinian schemes.

It turns out that, because the regions they bind are so small, developing a binding site for a regulatory protein is too easy. By chance, any particular six-nucleotide sequence should occur about once every four thousand or so nucleotides.[9] Given the enormous length of DNA, there is a great chance that a binding site might already be near a gene. What's more, the likelihood of having a site that matched five out of six positions—so that only one mutation would be needed to change the last position to make a perfect match—is even better. There should be one of those every few hundred nucleotides. Further, since there are so many regulatory proteins with different binding sites, potential binding sites that are one or two mutations away from binding some regulatory protein or other should be packed pretty much cheek by jowl in DNA.

Several studies have shown that is indeed the case. J. R. Stone and G. A. Wray have calculated that the likelihood of forming a new binding site for a given regulatory protein near a given gene, by random mutation in newborn organisms, is very high, about one in seventy.[10] Out of a million individuals in a generation, over ten thousand would have a shiny new site for any given control protein. In one person sick with malaria, for example, there would be ten billion new sites produced in a few days! In other words, an embarrassment of riches: There are so many potential binding sites that it's hard to conceive how they could be the chief factor determining whether a gene was turned on.[11]

You might object at this point that I seem to be impossible to please: Mutations are too rare, when we look at chloroquine resistance to malaria, but now they are too common, when we look at the theoretical possibilities for all these genetic switches. Here's the problem: So many kinds of switches are so common that, if they *were* the most important factor in determining whether a gene was turned on, the organism would be an incoherent mess. Instead of a fly or sea urchin or frog, a developmental program might at best produce a blob of tissue. In fact, as Stone and Wray explain, many, many other factors besides nucleotide sequence are required to be in place before a gene is activated.[12]

Remember the pyramid of gene switches, from egg-polarity genes to gap genes to pair-rule genes, and so on. The evo-devo hope is that such overarching control structures provide a way for zillions of simple mutations to toggle switches, making evolution somehow easier. Yet even with the new discoveries, a Darwinian path to the typical very complexly regulated eukaryotic gene would still have to be long and tortuous:

> The promoter regions of eukaryotic genes are complex and include approximately a dozen to several dozen transcription factor binding sites. The likelihood of a dozen binding sites evolving simultaneously without selection is infinitesimally small. . . . We envision instead that complex regulatory systems are the result of long and complex evolutionary histories involving stepwise assembly and turnover of binding sites.[13]

Modularity was supposed to make evolutionary changes simple—to smooth out a rugged evolutionary landscape. But, except for the unexpected complexity of genes and development, what exactly has changed? How have coherent changes been made easier?

In his review of *Endless Forms Most Beautiful* for *Nature*, University of Chicago evolutionary biologist Jerry Coyne is unimpressed by evo-devo claims.

> The evidence for the adaptive divergence of gene switches is still thin. The best case involves the loss of protective armour and spines in sticklebacks, both due to changes in regulatory elements. But these examples represent the loss of traits, rather than the origin of evolutionary novelties. Carroll also gives many cases of different expression patterns of *Hox* genes associated with the acquisition of new structures (such as limbs, insect wings and butterfly eyespots), but these observations are only correlations. One could even argue that they are trivial. . . . We now know that *Hox* genes and other transcription factors have many roles besides inducing body pattern, and their overall function in development—let alone in evolution—remains murky.[14]

In his book Carroll does not actually spell out how a novel structure would be built by evo-devo manipulations. Although he beautifully describes and illustrates fly embryology, he provides no specifics on how particular structures would evolve by random mutation and natural selection. In a typical passage, Carroll speculates about the evolution of insect wings.[15] He points to research showing that two certain control proteins found in wings are also found in crustacean gills, and concludes that the best explanation for this is that the organs are homologous—that is, the same body part in different forms in two different animals.

Like myriad biologists before him, Carroll confuses evidence for common descent with evidence for random mutation. Although, as he argues, the occurrence of the same control proteins in crustacean gills and insect wings may point to their common ancestry, it says absolutely nothing about how gills could be converted to wings by a Darwinian process.[16] In the same way, although one gene may flip the switch to trigger eye development, that tells us nothing about how unintelligent mechanisms could evolve an eye. Although studies of the genetics of embryology have unveiled breathtaking elegance and complexity, the ruminations of evo-devo proponents have—in my view—contributed little to the understanding of the evolution of complex structures.

THE FACILITATORS

Another new book by stellar researchers that trades heavily on the concept of modularity (which they call "compartmentation") is *The Plausibility of Life: Resolving Darwin's Dilemma*. Authors Marc Kirschner of Harvard and University of California–Berkeley's John Gerhart pick up where Sean Carroll left off (the jacket carries an appreciative blurb by Carroll and a handful of other high-powered scientists). They, too, recount the work of Monod and Jacob, the role of switches in controlling genes, Hox genes in *Drosophila*, and fruit fly development. They, too, emphasize that an animal's body can be subdivided into compartments by master regulatory genes, and that to a surprising extent the compartmentation is the same from fly to mammal. Unlike Carroll, however, they aim to fill in the blanks of Darwin's mechanism. They write that neither Darwin nor any of his contemporaries had a clue as to the underlying mechanisms needed to generate the variation from which nature could select. The tiny, random changes Darwin envisioned would have been grossly inadequate, think Kirschner and Gerhart. But evo-devo and modularity make random mutation more effective.

They christen their novel proposal "facilitated variation," signifying the idea that control genes make it relatively easy for organisms to vary in ways that might be evolutionarily helpful. If a complex system can be turned on by one simple trigger, and anything that pushes the button will work, then complexity is not necessarily an obstacle to Darwinism. Any input that flips the master regulatory gene for eye development in *Drosophila* will turn on the system and build an eye. Because links connecting genetic modules are "weak," they argue, systems and subsystems apparently can be disconnected, switched around, and reconnected pretty easily. Surely that would generate as much variation as Darwin could ask for. Dilemma resolved.

In fact Kirschner and Gerhart do not so much resolve Darwin's dilemma as invent a new and, in their view, better theory. Reviewers have not uniformly agreed.[17]

Kirschner and Gerhart, and indeed the entire evo-devo field, inadvertently do more to undermine Darwin than to save his theory. The first and most obvious concession made by evo-devo (tacitly or otherwise) is that profound, fundamental evolutionary questions had heretofore been utterly unexplained. The rise of multicellular animals, the appearance of novel processes and structures, not to mention novel cell types—none of those had been explained by Darwin's basic theory, even as elaborated by the "neo-Darwinian synthesis" of the mid twentieth century.

The next unwitting evo-devo point is even more striking: Basic features of life were totally unpredicted by Darwin's theory. In fact, *reasoning straightforwardly in terms of Darwin's theory* led badly astray even the most eminent evolutionary biologists, who reached conclusions completely opposite to biological reality. Consider the following examples:

- François Jacob wrote, "When I started in biology in the 1950s, the idea was that the molecules from one organism were very different from the molecules from another organism. For instance, cows had cow molecules and goats had goat molecules and snakes had snake molecules, and it was because they were made of cow molecules that a cow was a cow."[18] As Jacob ultimately learned, however, that was completely wrong.

- In the 1960s Ernst Mayr, an architect of the neo-Darwinian synthesis, confidently predicted on Darwinian grounds that "the search for homologous genes is quite futile," of which Sean Carroll notes, "The view was entirely incorrect."[19] In retrospect, it is astounding to realize that the strong molecular similarity of life, which Darwinists now routinely (and incorrectly) appropriate as support for their entire theory, was not anticipated by them. *They expected the opposite.*

- Mathematicians, too, were fooled. "Many theoreticians sought to explain how periodic patterns [such as fruit fly embryo segments] could be organized across large structures. While the maths and models are beautiful, none of this theory has been borne out by the discoveries of the last twenty years." "The continuing mistake is being seduced into believing that simple rules that can generate patterns on a computer screen are the rules that generate patterns in biology."[20]

- Writes Carroll, "The most stunning discovery of Evo Devo [that similar genes shape dissimilar animals] . . . was entirely unanticipated."[21] And "biologists were long misled" to think that simple legs were quite different from complex legs. "But it is wrong."[22]

- Kirschner and Gerhart are repeatedly surprised: They have a section entitled "The Surprising Conservation of Compartments."[23] And "It came as a surprise (if not a shock)" to find the same regulatory genes expressed in the heads of *Drosophila* and mammals. "Until that time, it was widely thought that the vertebrate head is entirely novel, the invention of our phylum."[24] According to Walter Gehring, the same goes for eyes. "This is an unexpected finding since the single lens eye of vertebrates was generally considered to have evolved independently of the compound eye of insects because these two eye types are morphologically completely different." [25]

Time and again, by intentionally reasoning about animal life on Darwinian principles, the best minds in science have been misled. They justifiably expected randomness and simplicity, but discovered depths of elegance, order, and complexity. As National Academy of Sciences president Bruce Alberts exclaimed, "We can walk and we can talk because the chemistry that makes life possible is much more elaborate and sophisticated than anything we students had ever considered." [26]

A third point is that, although it is polite and deferential, discontent with traditional Darwinism rumbles among many scientists who think most intently about evolutionary issues. As Smithsonian paleontologist Douglas Erwin wrote in his review of *The Plausibility of Life,* "Kirschner and Gerhart's book must be placed in the context of a number of other recent contributions to evolutionary thought, *all of which* argue that the current model of evolution is incomplete [emphasis added]." [27] Well, *is* the current model incomplete? "Is there reason to think that our view of evolution needs to change? *The answer is almost certainly yes*" [emphasis added], avows Erwin, quickly adding "although not, as the purveyors of creationism /intelligent design would have it, because the reality of evolution is under question." Apparently, by "the reality of evolution," Erwin means common descent, although he does not use this term.

In fact, some recent authors promoting modularity strongly insinuate that Darwin's theory as it has been understood by most biologists of the past century and a half *could not* account for major features of life. Only now is it credible. If the most recent findings were not correct, they say, Darwinism would be forlorn. Toward the end of their book Kirschner and Gerhart coyly ask;

> Can evolution be imagined without facilitated variation? What capacity to evolve would a hypothetical organism have if it did *not* have facilitated variation? If animals did not use and reuse conserved processes, they would, we think, have to evolve by way of total novelty—completely new components, processes, development, and functions for each new trait. Under these circumstances

the demands for "creative mutation" would be extremely high, and the generation of variation might draw on everything in the phenotype and genotype.[28]

Their clear implication is that without facilitated variation—their own brand-new proposal—Darwinism would fail.

As a computer scientist interested in evolutionary algorithms, University of Southampton lecturer Richard Watson comes at the topic from a different angle, but he arrives at the same conclusion as Kirschner and Gerhart. In *Compositional Evolution*, Watson lays it on the line:

> In computer science we recognize the algorithmic principle described by Darwin—the linear accumulation of small changes through random variation and selection—as *hill climbing*, more specifically *random mutation hill climbing*. However, we also recognize that hill climbing is the simplest possible form of optimization and *is known to work well only on a limited class of problems* [emphasis added to the last sentence].[29]

Those problems include very simple ones that can be solved by changing just one or a few variables—as in the evolution of drug resistance or the resistance of humans to malaria. "Darwin's masterful contribution was to show that there was *some* principle of optimization that could be implemented in biological systems," allows Watson—just not the right one for complex systems. Watson proposes his new idea of "compositional evolution," which boils down to more modularity. Without compositional evolution, implies Watson, evolution by unintelligent processes would be a no-go.

In sum, the new evolutionary writings have unintentionally done much to damage Darwin, but have not offered convincing alternatives to replace him.

IT ONLY GETS WORSE

Let's acknowledge that genetics has yielded yet more terrific (and totally unanticipated) evidence of common descent. Has evo-devo produced a new way for random mutation to explain basic features of animal life? No, exactly the opposite. It's not hard to see why, more than twenty years after the first animal control proteins were sequenced, evolutionary biologists are still utterly unable to give a concrete account of how to explain the unintelligent evolution of animal forms.

In Chapter 7 we encountered unanticipated bottom up–top down construction in systems that build cellular machinery such as the cilium. In retrospect, we realized that the need for specific systems to construct a cilium, as well as for intricate genetic control programs to coordinate the construction, greatly complicates the task of explaining them. The control systems are *a further layer of complexity*—on top of the complexity of the finished systems themselves—which we in our innocence had not fathomed would be required. The need for control systems does not make the task of Darwinian explanation easier; it makes it far worse.

In the same way as for molecular machinery, in the past several decades developmental biology has unexpectedly discovered the need for careful, bottom up–top down planning in the construction of the entire animal. As Eric Davidson trenchantly noted, "Development imposes *extreme* regulatory demands. . . . *All* of the structural characters of the edifice, from its overall form to *minute aspects* that determine its local functionalities such as placement of wiring and windows, *must be specified* in the architect's blueprints" (my emphasis). As with molecular machinery, the elaborate assembly control instructions for whole animals are a *further layer of complexity*, beyond the complexity of the animal's anatomy itself. The inadequacy of Darwinism to account for the intricacies of animal development has not been lessened by recent discoveries; it has been greatly exacerbated.

ALL THINGS CONSIDERED

Even though the castaway of Chapter 8 didn't have hard estimates of probabilities, in light of his experience of nature and his sure knowledge of the design of the wrecked ship, he confidently judged that the neatly piled square of stones and other island features were purposefully designed, rather than the result of some bizarre accident such as a lightning strike. Similarly, although hard numbers are difficult for us to come by, in light of our knowledge of the design of spectacular molecular systems such as the cilium and our experience of nature (particularly our experience with the havoc wreaked by random mutation—even when it "helps"), we can confidently judge that the kind of coherent, multistep control system that Davidson's observation indicated was demanded to build an animal body was purposely designed.

But how deep does that design extend? There are many distinct animal forms, which biologists have long placed into hierarchical categories such as phyla, classes, and orders. We must remember that randomness does occur and can explain some aspects of all areas of life. So, based on developmental biology and our new knowledge of life's molecules, can we draw a reasonable, tentative line between Darwin and design in animal evolution? Does design stop at, say, the level of phyla? Or classes? For example, given a generic animal in the distant past with twofold, bilateral symmetry, is it biologically reasonable to think that at that point the rest of the animal world could evolve by random mutation? Or not? Again, we have to keep in mind that few pertinent, quantitative experiments directly applicable to that question have been done. What's more, further lab work will almost certainly uncover much greater complexity in animal development and other relevant facts, so our appraisal will have to be revised as more information comes in. Nonetheless, there are enough data already in hand to form a reasoned estimate.

To prepare to locate a provisional edge of animal evolution, let's consider several important factors. When we pass from considering single-celled creatures to multicelled animals, two big things change, in opposite directions. First, we find that animals already are endowed with a passel of toolbox components such as Hox genes to play with (which is where evo-devo musings generally start). That just might open up random-evolutionary possibilities.

(As an aside, it is fascinating to note that the appearance of Hox toolbox components seems to have significantly predated the appearance of new animal forms. As Sean Carroll remarks:

The surprising message from Evo Devo is that all of the genes for building large, complex animal bodies long predated the appearance of those bodies in the Cambrian Explosion. The genetic potential was in place for at least 50 million years, and probably a fair bit longer, before large, complex forms emerged.[30]

Another surprise to Darwinists! To an intelligent design proponent such as myself, this is a tantalizing hint that parts were moving into place over geological time for the subsequent, purposeful, planned emergence of intelligent life.)

But second, population sizes plummet, which greatly restricts Darwinian possibilities. No multicelled species can match the sheer population numbers that bacteria reach. When we consider animals, we now have many, many fewer than 10^{40} organisms—the number of bacterial cells that have likely existed on the earth since it formed. As pointed out earlier, the number of malarial parasites produced in a single year is likely a hundred times greater than the number of all the mammals that have ever lived on earth in the past two hundred million years.[31]

As the population sizes associated with multicellular organisms drop, they begin to fall out of the "couch potato" evolutionary class into the "frail old man" class. In other words, unlike single-celled organisms, larger multicelled animals can no longer be expected to jump more than one missing mutational step, simply because they have fewer chances to generate beneficial mutations. As a rule, each and every mutation—each nucleotide or amino acid change—along the path to a new feature would have to be either beneficial or at the very least not harmful. To reiterate Allen Orr's conclusion, "Given realistically low mutation rates, double mutants will be so rare that adaptation is essentially constrained to surveying—and substituting—one-mutational step neighbors."[32]

How does that affect our estimation? A reasonable, informed person would find it hard to disagree with Stone and Wray's expectation, "as with amino acid substitutions within coding regions of genes, we predict that in many cases the consequences of a new binding site appearing within a promoter will be either detrimental or neutral; only in rare cases will it be beneficial."[33] So here is the key judgment: It seems a reasonable approximation to treat changes in switch regions, regulatory proteins, and so on, roughly the same as changes in protein-binding sites. That is, if some new control mechanism requires several coherent steps to set it up, for example two or three control proteins acting in concert, then it is reasonable to consider that as roughly equivalent to several proteins binding to each other in a useful multiprotein complex, and to rule out random mutation as an explanation for it.

Why does the fact of multiple coherent steps matter? In Chapter 3 we saw that resistance of malaria to chloroquine was found in only one in a hundred billion billion organisms (10^{20}—a CCC) because it required skipping an evolutionary step. In Chapter 7 I argued that three different proteins (two new binding sites) forming a specific complex was beyond the molecular edge of evolution, because it was a double CCC, 10^{40}. The likelihood of the event was so low it would not be expected to occur in the history of the earth, because an organism would have to jump a number of evolutionary steps. Here, with many fewer organisms available, the argument is that forming a new control mechanism for some feature of animal development involving about three or more different kinds of proteins or switches is also a reasonable place to draw the edge of evolution for animal form, because, again, evolutionary steps would probably have to be skipped.

Admittedly, this is a fuzzy estimate—necessarily so, because our current data are limited. Nonetheless, the uncertainty shouldn't deter us from reaching at least some reasonably firm judgments, because some major control mechanisms uncovered so far are well beyond this measure.

DEEPER AND DEEPER

First, let's consider a control mechanism that is known to be very complex—one that showcases the sense of Eric Davidson's exclamation that "development imposes extreme regulatory demands." A recent special issue of the *Proceedings of the National Academy of Sciences* explored "genetic regulatory networks"; that is, the control machinery that is necessary to build animal bodies. As the editors Michael Levine and Eric Davidson explain:

Endomesoderm Specification to 30 Hours

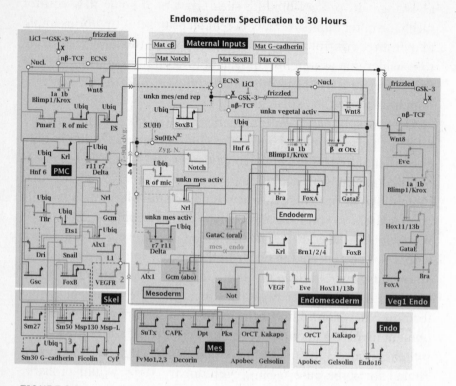

FIGURE 9.3

Schematic drawing of a developmental gene regulatory network for sea urchin endomesoderm. The network is strongly evocative of a complex electrical or computer-logic circuit. The figure is reproduced from http://sugp.caltech.edu/endomes/, courtesy of Eric Davidson, who wished to have noted that "Permission for use of this figure is not to be construed to indicate the agreement of its authors with the overall thesis" of this book.

Gene regulatory networks (GRNs) are *logic maps* [emphasis added] that state in detail the inputs into each *cis*-regulatory module, so that one can see how a given gene is fired off at a given time and place. . . . The architecture reveals features that can never be appreciated at any other level of analysis but that turn out to embody distinguishing and deeply significant properties of each control system. These properties are composed of linkages of multiple genes that together perform specific operations, such as positive feedback loops, which drive stable circuits of cell differentiation.[34]

Figure 9.3 is an illustration of the genetic regulatory system that turns on the genes that control the construction of a tissue called the endomesoderm in sea urchins. Notice the obvious, impressive coherence of the drawing. The figure is intended to be strikingly reminiscent of a complex electronic or computer-logic circuit, because in essence that is what genetic circuits are. The system contains a core of six genes that code for master regulatory proteins that eventually switch on scores of proteins that boast many more DNA switches, very far beyond the criterion of three proteins or switches. We can thus conclude this system is well beyond the edge of evolution. It was very likely purposely designed.

Eric Davidson and Douglas Erwin describe the core of the control system for sea urchin endomesoderm as a genetic regulatory network "kernel," the most basic type of regulatory network now known.[35] Kernels, they say, have a number of properties, including: 1) they "specify the spatial domain of an embryo in which a given body part will form"; and 2) "interference with expression of any one kernel gene will destroy kernel function altogether"—in other words, they are irreducibly complex. If all the genes are necessary for kernel function, it would have required many coherent evolutionary steps to set up. Kernels in general can be expected to have a degree of complexity similar to that for sea urchin endomesoderm, so we can infer that other kernels also were designed.

Animals are divided into a number of groups according to their general "body plan." For example, one group of animals, chordates (which includes vertebrates like us), have a nerve cord arranged in the back of their bodies, whereas arthropods, the group that includes insects and crustaceans, have a nerve cord in the front. Biologists count dozens of fundamentally different body plans. Types of animals that have the same body plan are generally grouped together in the same phylum, which is the biological classification right under kingdom (kingdom divides organisms into bacteria, plants, animals, and a few other categories).

Now, since kernels "specify the spatial domain of an embryo," kernels must designate different body plans. Although there are many gaps in our knowledge, as Davidson and Erwin remark, "There are a number of additional examples for which there is persuasive evidence for the existence of [genetic regulatory network] kernels awaiting discovery of the direct genomic regulatory code. Prospective examples include kernels common to all members of a given phylum or superphylum." Therefore, because a crucial element of body plan development—the kernel—requires design, it is reasonable to consider body plans in general to be designed. So we can further conclude that design extends into life at least as far as animal phyla.

Of course, animals from different phyla share many features. For example, all animals are eukaryotes, and thus have cells with nuclei and a molecular skeleton. Nonetheless, recall the bicycle/motorcycle example I mentioned at the beginning of this chapter. Although the two-wheeled vehicles share some parts, it's reasonable to view a motorcycle as a separate, integrated design. Following that reasoning, it seems likely that different phyla represent separate, integrated designs.

Does design extend further into life than phyla? Yes, very likely. A hallmark of animal development is the differentiation of cells into different types, such as muscle cells, skin cells, and retinal cells. Because of the medical importance of the immune system, excellent work has been done on how one type of immune cell, called a "B cell," is formed. In the special issue of the *Proceedings of the National Academy of Sciences* that featured genetic regulatory networks, an article summarized B cell differentiation. Although work is tentative and is continuing, the number of protein factors known to be involved in the gene regulatory network for B cell differentiation is similar to the number involved in the endomesoderm kernel (about ten).[36] The authors comment:

> The B cell developmental pathway represents a leading system for the analysis of regulatory circuits that orchestrate cell fate specification and commitment. . . . [T]he proposed circuit architecture is foreshadowing design principles that include transient signaling inputs, self-sustaining positive feedback loops, and crossantagonism among alternate cell fate determinants.

Thus, because of its coherence and the number of its components (well beyond our criterion of three), it's reasonable to think the system to specify B cell differentiation was also designed. B cells don't occur in invertebrates; they are found only in vertebrates. Based on just this one particular example, then, it appears that design extends into the phylum Chordata, past the divide between invertebrates and vertebrates, which is the level of subphylum.

The work that goes into elucidating gene regulatory networks is enormous. At this point the B cell is one of the very few cell types where much is known about the gene networks that control its differentiation. However, if we assume that the B cell regulatory network is typical of what is needed to specify a cell type, we can conclude that design is required for new cell types in general. That will move us further along. Vertebrate classes differ in the number of cell types they have. Although amphibians have about 150 cell types and birds about 200, mammals have about 250.[37] So, again keeping in mind the limitations of the data, because different classes of vertebrates need different numbers of cell types, we can tentatively conclude that design extends past vertebrates in general and into the major classes of vertebrates—amphibians, reptiles, fish, birds, and mammals.

Does design extend even further into life, into the orders or even families of vertebrate classes? To such creatures as bats, whales, and giraffes? Because "all of the structural characters of the edifice, from its overall form to minute aspects that determine its local functionalities . . . must be specified in the architect's blueprints,"[38] I would guess the answer is almost certainly yes. But at this point our reliable molecular data run out, so a reasonably firm answer will have to await further research. Given the pace of modern science, we shouldn't have to wait too long.

BRACKETING THE EDGE

Does the reasoning above comport with what's known from observational data? Yes. Let's divide the answer into negative results and positive results. First, briefly, the negative. Of the many human genetic changes wrought by the struggle with malaria in the past ten thousand years, a few occur in regulatory regions. But nothing is built—single genes are simply shut down or deregulated; there are no new genetic regulatory systems formed. The same kind of small, incoherent changes we see in humans occur in other animal species, too. In a billion rats in the past fifty years, evo-devo theorists might expect many new regulatory regions; none seem to have helped against warfarin. In trillions of Antarctic notothenioid fish in the past ten million years, no new regulatory regions seem to have helped much in the fight against freezing water—only changes in protein sequences do. In the laboratory, the fruit fly has been studied in large numbers for over a century. Although its existing genetic control systems have been subjected to all manner of experimental insults, resulting in some bizarre birth defects, during that time no new, helpful, developmental-control programs have appeared.

The malarial parasite is a single-celled organism, so of course it does not need a body plan. Nonetheless, during its life cycle it changes between several distinct forms, which can be considered as akin to cell types. Yet in a hundred billion billion chances, no new cell forms or regulatory systems have been reported. What greater numbers of malaria can't do, lesser numbers of large animals can't do either. In other words, as expected, there is no evidence from our best evolutionary studies that random mutation leads to gene regulatory networks of the complexity of cell differentiation—that is, class-level biological distinctions.

On the positive side, some terrific work has been done in recent years yielding some persuasive evidence that random changes in existing control networks can helpfully affect animal form at the species level. One analysis that will warm the heart of any pet owner was an investigation of possible molecular reasons for the differences between breeds of dogs. A recent study adduced evidence that changes in some dog Hox genes, where one or several amino acid codons are repeated a varying number of times, are correlated with some differences in bone structure among breeds.[39]

A few other studies were highlighted in a recent issue of *Science* that designated "Evolution in Action" as the "Breakthrough of the Year."[40] One looked at the differences between two varieties of stickleback fish.[41] It concluded that the ancestral form usually found in oceans, which has more bony armor in its body and three bony spines sticking up from its back, has given rise several times independently over the past few million years to a form usually found in fresh water, which has much less armor and fewer spines, probably due to mutations in certain control regions. Another study from the group led by Sean Carroll showed that males of a certain species in the genus *Drosophila* in the past 15 million years have gained a spot of color on their wings. The reason is that the gene for a pigmentation protein called Yellow protein (which actually produces dark pigmentation) has gained a new switch sequence for a particular regulatory protein.[42] This result is important because it shows random mutation not only breaks switches, but occasionally makes new ones, too, just as it occasionally makes proteins with new functions such as the antifreeze protein of notothenioid fish.

These studies are great reminders that random mutation and natural selection can account for many relatively minor changes in life—not only changes in invisible metabolic pathways like antibiotic resistance in rats or malaria, but also changes in the appearance of animals. The different sizes and shapes of dogs, the patterns of coloration of insect wings, and more can very likely be attributed to Darwinian processes affecting gene switches.

Combining the reasoning from the past several sections, then, we can conclude that animal design probably extends into life at least as far as vertebrate classes, maybe deeper, and that random mutation likely explains differences at least up to the species level, perhaps somewhat beyond. Somewhere between the level of vertebrate species and class lies the organismal edge of Darwinian evolution.

DARWIN AMONG THE SPANDRELS

So, given these results, if you are willing to consider the possibility of intelligent design, how should you view the relationship of Darwin to design? Although I'm sure they would disapprove, I think a felicitous image can be borrowed from a well-known paper by the late evolutionary biologist Stephen J. Gould and the Harvard geneticist Richard Lewontin entitled "The Spandrels of San Marco."[43] A spandrel is an architectural term that designates the "tapering triangular spaces formed by the intersection of two rounded arches at right angles." As Gould later recounted, they co-opted the term "to designate the class of forms and spaces that arise as necessary byproducts of another decision in design, and not as adaptations for direct utility in themselves."[44] In other words, the joint between two designed structures has to look like something, but it's a mistake to think the seam was necessarily intended for itself.

As an example of an architectural spandrel, Gould and Lewontin pointed to the "great central dome of St. Mark's Cathedral in Venice." Each of the four tapering spaces where rounded arches intersect in the cathedral is decorated with elaborate art, including a painting of one of the four evangelists. The painting fits so harmoniously with the cathedral, they wrote, that if you didn't know better, you might think the whole structure was built just to give a space for the decorations. Yet the paintings, fitting as they may be, are merely filling an open niche. For similar reasons, the authors also pointed to the fan vaulted ceiling of King's College Chapel of Cambridge University, where some open space along the midline is decorated with the Tudor rose. "In a sense," they wrote, "this design represents an 'adaptation,' but the architectural constraint is clearly primary."

And so it is between design and Darwin in life. The major architectural features of life—molecular machinery, cells, genetic circuitry, and probably more—are purposely designed. But the architectural constraints leave spandrels that can be filled with Darwinian adaptations. Of course, Darwinian processes would not produce anything so coherent as the paintings of the four evangelists. Random mutation and natural selection ornament biological spandrels more in the drip-painting style of the abstract American artist Jackson Pollock. The myriad gorgeous color patterns of animals—butterfly wings, tiger stripes, bright tropical fish—are some examples of Darwin among the spandrels.

Figure 9.4
A spandrel formed by two designed arches. (Drawing by Celeste Behe.)

Darwin decorates the spandrels. The cathedral is designed.

10

ALL THE WORLD'S A STAGE

CONSILIENCE

"Consilience" is an old-fashioned synonym for concurrence or coherence. When results from separate scientific disciplines all point in the same direction, we can be far more confident of the conclusion. About a decade ago the noted biologist E. O. Wilson wrote a book titled *Consilience*. Wilson argued that ideas from Darwinian evolutionary biology can illuminate other areas of knowledge, such as environmental policy, social science, and even the humanities. Because of this, he thinks he sees a consilience of results that supports what is variously called scientism, reductionism, or materialism—in other words, the view that the entire universe from the Big Bang to the Bolshoi Ballet can be explained by the random, unguided playing out of natural laws.

I think Wilson has it exactly backward. Rather than supporting randomness, a consilience of relatively recent results from various branches of physical science—physics, astronomy, chemistry, geology, molecular biology—actually points insistently toward purposeful design in the universe. In each case the results were unexpected and surprising. Merely intriguing when considered in isolation, when taken together the results from the disparate disciplines strongly reinforce each other. They paint a vivid picture of a universe in which design extends from the very foundations of nature deeply into life.

MINNESOTA FATS

Here's a brief analogy to help think about the new consilience. Suppose in a small room you found a pool table, with all the pool balls held in one side pocket. Nothing much remarkable about that, you tell yourself. Now suppose you later discovered a videotape from an overhead camera, showing how the balls arrived in the pocket. As the tape begins, all the numbered pool balls are motionless, scattered on the table apparently at random. Then, in slow motion from one corner, the cue ball appears (you can't see the cue stick or shooter—they're off-camera). The cue ball hits a numbered ball, then another, which hits several others. After bouncing around a short while, all the balls line up and roll neatly, one after another, in numerical order, into the side pocket.

Even though you didn't see what happened before the start of the film or off-camera, you would be certain it was a trick shot. No random cue stroke, that. It was set up—designed. The shot must have taken into account not only general laws of physics (conservation of momentum, friction, and so forth) but special conditions (the size of the table and mass of the balls) as well as minute details (the exact initial placement of the balls and angles of impact). Whoever set up the trick not only took care to select appropriate general conditions, including a smooth pool table, but also paid close attention to the smallest details necessary to make the trick work.

The pool table is our universe, and the consequence of all the balls in the side pocket is life on earth. The initial, static, naive view—the balls already in the pocket, where you first spot them—is akin to nineteenth-century science's view of the universe. The jaw-dropping dynamic view given by the videotape is analogous to what modern science has discovered.

FINELY TUNED *LAWS*

In the second half of the nineteenth century the universe seemed pretty dull. It was thought to be eternal and largely unchanging, composed of relatively simple matter, obeying a few rules such as Newton's law of gravity. Such a cosmos could have been mistaken for a background of boring wallpaper. In the most spectacularly wrong consensus in the history of science, in the words of two historians, "At the end of the nineteenth century there was a general feeling that, with Maxwell's and Newton's equations firmly established, everything else would be merely a matter of detail, a question of dotting the i's and crossing the t's of science."[1] As Yogi Berra observed, it's tough to make predictions, especially about the future. Soon Einstein proposed his theory of relativity; quantum mechanics swept through physics; the atom was shown to be divisible—into protons, neutrons, and much more. Like the cell, which was also thought to be simple, the universe became more complex the more it was studied.

First the wallpaper was revealed to be full of strange details. Then the very size and shape of the room began to change. In 1929 the astronomer Edwin Hubble measured light coming from distant galaxies. He was startled to see that the wavelength of the light was somewhat longer than it should have been. The so-called "redshift" of starlight is similar to what happens when a speeding train, blowing its whistle, passes a person standing by the tracks, who hears the pitch of the whistle change from higher to lower as the train recedes. Hubble interpreted the redshift of starlight to mean that galaxies are rapidly receding from the earth and from each other, as if in the aftermath of a huge explosion. This was the beginning of the Big Bang theory, and the end of the humdrum, eternal, unchanging universe.

In the second half of the twentieth century physics advanced by leaps and bounds. More and more subatomic particles and forces were discovered, more and more measurements and computer calculations accumulated. In the mid-1970s a physicist named Brandon Carter paused to think about the new data from the viewpoint of what's needed for life. In a paper entitled "Large Number Coincidences and the Anthropic Principle in Cosmology," Carter pointed out that if any of a number of the multiple laws and constants that physics had discovered in the twentieth century had been a tiny bit different, the universe would be utterly unsuitable for life.[2] In other words, the very same cosmos that appeared so bland just a hundred years ago now is known to be balanced on a knife edge, with numerous factors arranged just-so, to permit life.

Since Carter's seminal paper many commentators have remarked on the astounding "fine-tuning" in physics. Consider this oft-quoted passage from the physicist Paul Davies:

> The numerical values that nature has assigned to the fundamental constants, such as the charge on the electron, the mass of the proton, and the Newtonian gravitational constant, may be mysterious, but they are crucially relevant to the structure of the universe that we perceive. As more and more physical systems, from nuclei to galaxies, have become better understood, scientists have begun to realize that many characteristics of these systems are remarkably sensitive to the precise values of the fundamental constants. Had nature opted for a slightly different set of numbers, the world would be a very different place. Probably we would not be here to see it.
>
> More intriguing still, certain structures, such as solar-type stars, depend for their characteristic features on wildly improbable numerical accidents that combine together fundamental constants from distinct branches of physics. And when one goes on to study cosmology—the overall structure and evolution of the universe—incredulity mounts. Recent discoveries about the primeval cosmos oblige us to accept that the expanding universe has been set up in its motion with a cooperation of astonishing precision.[3]

Similarly, the Cambridge University physicist Stephen Hawking remarks:

> The laws of science, as we know them at present, contain many fundamental numbers, like the size of the electric charge of the electron and the ratio of the masses of the proton and electron. . . . The remarkable fact is that the values of these numbers seem to have

been very finely adjusted to make possible the development of life. For example, if the electric charge of the electron had been only slightly different, stars either would have been unable to burn hydrogen and helium, or else they would not have exploded [which allows elements necessary for life to be scattered]. . . . It seems clear that there are relatively few ranges of values for the numbers that would allow the development of any form of intelligent life.[4]

So what are we to make of the flabbergasting fact that the laws of the universe seem set up for our benefit, that "the universe in some sense must have known that we were coming"?[5] There are really just two logical responses to anthropic features: 1) We are phenomenally lucky, or 2) our universe was intentionally designed by an intelligent agent. Over the next few sections I'll consider the design explanation for anthropic features of our universe, and extend the argument far into biology.

FINELY TUNED *PROPERTIES*

Many, many other factors aside from the laws of physics need to be just right before one gets a planet that can nurture intelligent life. For example, not only does the physics of elementary particles have to be right, so do the physics and chemistry of molecules. In other words, having the right value of, say, the charge on an electron so that molecules are stable is just the first, tiny step. The molecules of life have to have other, useful properties, beyond the basics.

The most famous example is water. Water is such a familiar liquid that most people don't give it much thought. Yet scientists know it's unique, and that life without water is virtually unimaginable. Almost all other liquids contract when they freeze. Water expands. Although that seems like a trivial feature, it's critical for life. If water contracted on freezing, ice would be denser than water, and would sink to the bottom of a lake or ocean, away from warming sunlight. Virtually all the water on earth would likely be frozen solid, unavailable for life. As the geneticist Michael Denton points out in *Nature's Destiny*,[6] water also has many other properties that suit it to be life's liquid. Just one further example is that water can dissolve a wide range of substances, such as salts and sugars; very few other liquids can. Of those few other liquids, many are either strongly acidic or strongly basic, or are otherwise unsuited for life. Denton makes a strong case that, in addition to water, many other elements and simple molecules—carbon, oxygen, carbon dioxide, metals, and many more—are as necessary for life as the fundamental constants and laws of nature. So the "anthropic coincidences" needed for life in this universe extend beyond the basic physical laws and constants, well into chemistry.[7]

What is the explanation for such a remarkable, unexpected pattern? One great virtue of the design hypothesis is that, without making additional assumptions, it supports all the further anthropic coincidences found more recently in physics and chemistry. After all, an agent who can actually choose and establish the basic laws and constants of the universe is, to say the least, likely to be immensely intelligent and powerful, and so have the ability to further fine-tune nature as necessary. What's more, if the agent evinces an interest in life, as reflected in biofriendly general laws, then we should expect it would take whatever additional steps would be necessary to achieve its goal of life. The fact that we have discovered life required much more fine-tuning than first supposed fits easily into the design explanation.

If one admits the possibility of a being who can fine-tune general laws, then there can be no principled objection to ascribing other fine-tuned features of nature to purposeful design. In its 1999 booklet *Science and Creationism*, the National Academy of Sciences (or at least a committee writing in its name) penned the following lines:

> Many religious persons, including many scientists, hold that *God created the universe and the various processes* driving physical and biological evolution and that these processes then resulted in the creation of galaxies, our solar system, and life on Earth. This belief, which sometimes is termed "theistic evolution," *is not in disagreement with scientific explanations of evolution*. Indeed, *it reflects the remarkable and inspiring character of the physical universe* revealed by cosmology, paleontology, molecular biology, and many other scientific disciplines [emphasis added].[8]

It seems to me likely in this passage the committee was simply trying to make a reassuring gesture toward religious folks, while simultaneously doing what it could to steer the public's religious beliefs toward those that cause the least trouble for Darwin's theory. But the committee did not think through the implications of its words. Because if there is indeed a real being who could actually create the universe and its laws, as the committee allows, and if that explanation reflects (that is, is evidentially supported by) "the physical universe revealed by cosmology" and other scientific disciplines, what would stop the being from affecting the universe in other ways if it chose to do so? Would this being that created the universe and its laws have to ask permission of the National Academy to otherwise affect nature? Of course not. Whether it affected the universe in additional ways would be a matter for the evidence—not the committee—to decide. The bottom line is that, if one allows that a being external to the universe could affect its laws, there is no principled reason to rule out a priori more extensive interaction as well.

The consilience of fine-tuning in physics and chemistry reinforces our confidence in design. It's reasonable to conclude not only that the universe is designed, but that the design extends well beyond general laws, at least down into particularities of the physics and chemistry of certain molecules.

FINELY TUNED *DETAILS*

In the past few decades it has been gradually realized that anthropic coincidences extend well beyond even the physical and chemical properties of particular compounds such as water. Anthropic coincidences now include a long list of what can only be termed *details*. And pretty minute details at that.

In *Rare Earth: Why Complex Life Is Uncommon in the Universe* geologist Peter Ward and astronomer Donald Brownlee present a powerful argument that, as the title suggests, planets like ours may be exceedingly rare in the universe, in fact so rare that, although the authors think other planets may sport primitive bacterial life, ours may be the only planet able to support intelligent life. If that's the case, then, like a pool shark setting up a trick shot, the agent who set up the universe for intelligent life would have had to pay attention to all the details needed to produce an appropriate planet.

(Ward and Brownlee themselves think the earth was lucky, not designed: Given the size of the universe, getting one or a relative handful of planets physically like earth may be possible just by chance. However, that thinking overlooks problems with the origin of life and evolution by random mutation. If the subsequent evolution of intelligent life—even on a suitable planet—is itself enormously improbable, the "lucky" line of reasoning breaks down. In that case there would be just one or a few earthlike planets, but terrible odds against any of them developing intelligent life. We would have no good reason to expect life to develop on a single earth.)

Not only do the laws and chemical properties have to be right, but planets have to form in the right location with the right mix of ingredients. That's very difficult to do. If the earth were a bit closer to the sun, it would be too hot to support intelligent life. Venus is 68 million miles from the sun. The earth is 93 million miles away. On Venus temperatures are high enough to melt lead. If the earth were a little farther from the sun, water would freeze. Daytime temperatures on Mars, which is about 140 million miles from the sun, average about —80°F. The so-called "Goldilocks effect," where a planet can't be located where it's too hot or too cold, has led to the notion of a "habitable zone"—a narrow region in a solar system that has more of the necessary, but still far from sufficient, conditions for life.

The concept of a habitable zone applies to galaxies, too. It turns out that large swaths of a typical galaxy are quite hostile to life. Regions that are either too close to a galaxy's center or situated in a band of a spiral galaxy get fried by high doses of X-rays emanating from colliding stars and supernovas. Regions too far from the center of a galaxy have a different set of problems, as Ward and Brownlee explain.

> The outer region of the galactic habitable zone is defined by the elemental composition of the galaxy. In the outermost reaches of the galaxy, the concentration of heavy elements is lower because the rate of star formation—and thus of element formation—is lower. Outward from the centers of galaxies, the relative abun-

dance of elements heavier than helium declines. The abundance of heavy elements is probably too low to form terrestrial planets as large as earth. . . . [O]ur planet has a solid/liquid metal core that includes some radioactive material giving off heat. Both attributes seem to be necessary to the development of animal life: The metal core produces a magnetic field that protects the surface of the planet from radiation from space, and the radioactive heat from the core, mantle and crust fuels plate tectonics, which in our view is also necessary for maintaining animal life on the planet. No planet such as Earth can exist in the outer regions of the galaxy.[9]

Decades ago the late astronomer Carl Sagan derided the Earth's location as a galactic backwater. But with the progress of science we now see that a planet suitable for life can't be too close to the center of things. Far from being a backwater, Earth's location is ideal for complex life.

A planet in the right region of a solar system, in the right region of a galaxy, in a universe with the right kinds of laws to produce chemicals with the right kinds of properties—this is all necessary for life, but still very far from sufficient. The planet itself has to be not too big and not too small, with enough but not too much water, the right kind of minerals in the right places, a core active enough to power plate tectonics but not so active as to blow everything apart, and much, much more. Some of these factors considered in isolation may be less improbable, others more improbable, but all are critical. If any one of them were missing, intelligent life would be precluded.

FINELY TUNED *EVENTS*

To get a better feel for the extreme fine-tuning required to produce intelligent life, consider the critical role played in the formation of the earth by a unique *event*—that is, by a singular occurrence that is not simply a "law," or a "property," or even a "detail." Instead, like a cluster of pool balls bouncing just so, it results from the dynamic interplay of multiple, apparently unrelated factors.

Consider the entwined origin of the earth and moon. The question of how the moon originated puzzled astronomers for centuries. In recent years a radical proposal has emerged as the most likely contender. Billions of years ago, as the nascent earth itself was forming, a large body roughly the size of Mars struck our then-undersized planet. It proved to be a spectacularly accurate, glancing blow. The tremendously energetic, not-quite-head-on impact caused the two bodies to melt. The dense molten metal cores of the two orbs combined and sank into the interior of the now-larger earth, while chunks of the rocky crust of both were ejected into space, later to coalesce into a stable, orbiting moon. As Ward and Brownlee remark, "To produce such a massive moon, the impacting body had to be the right size, it had to impact the right point on Earth, and the impact had to have occurred at just the right time in the Earth's growth process."[10]

The moon that resulted from that seemingly serendipitous, unique event contributes in a variety of critical ways to making our planet livable, for example by stabilizing the earth's tilt—that is, the angle formed by earth's axis of rotation compared to its orbital plane—which allows earth's climate to avoid extreme, life-killing fluctuations. In short, without that singular collision in space when the solar system was young, as well as the laundry list of necessary conditions that preceded it and the plethora of happy results that flowed from it, earth would be uninhabitable, no matter that it was otherwise in the right location.

The bottom line is that the "fine-tuning" of our universe for life is not at all just a matter of the basic laws and constants of physics. Fine-tuning reaches deeply into ostensibly small details of the history of our solar system and planet, and includes unique, dynamic events. Without attention to such small details, mere fine-tuning of general physical laws would be futile. The strong nuclear force might be perfect, the charge on the electron just right, but if the end result of the undirected playing out of general laws were a lifeless asteroid where the earth should have been, why bother?

If there really does exist an agent who tuned the general laws of nature with the goal of producing intelligent life, then it's reasonable to think the agent would have taken whatever further steps were necessary to achieve its goal. And, to science's great surprise, in the past century it has discerned an ever-lengthening list of essential steps. If the design hypothesis is a leading contender as an explanation for the fine-tuning of the laws of the universe, then by the same reasoning it also must be regarded as a leading explanation for the finer physical, chemical, and astronomical details and events that make life possible. Arranging for a happy collision between two astronomical bodies is not obviously more difficult than arranging the right values for the fundamental constants of nature, so there is no principled reason to allow the possibility of design in the one case but withhold it in the other.

THE ORIGIN OF LIFE AS A
FINELY TUNED EVENT

The laws and constants of the universe are finely tuned to allow life. So, too, are the physical and chemical properties of elements such as carbon and simple compounds such as water. So, too, is earth's location in the galaxy and solar system. So, too, are details of the earth's size, composition, and history.

So, too, as the geneticist Michael Denton has forcefully argued in *Nature's Destiny,* are the more complex categories of the molecules of life. The physical and chemical properties of DNA, protein, lipids, and other substances are superbly fit for the roles they play in the cell. No other kinds of molecules are known that could plausibly fill those roles. What's more, shows Denton, the complex molecules of life harmonize in multiple ways with many levels of nature. For example, the strength of the electric charge allows both the strong (covalent) and the weak (noncovalent) chemical bonds necessary for proteins to work. Denton makes the case that one would be hard pressed to find a category of biomolecule or basic feature of life that *isn't* finely tuned.

Here I want to extend the rubric of "fine-tuning" even further to embrace the origin of life. Just as astronomers for decades beat their collective heads against the wall trying to envision a comparatively orderly, lawlike scenario for the origin of the moon, only to come up empty, so, too, have origin-of-life researchers for decades beat their heads against a wall trying to come up with a comparatively lawlike scenario for the origin of life, only to come up empty. The current model for the origin of the moon is not lawlike in the least, no more than a trick pool shot is lawlike. In isolation it can only be described as utterly random. But in context, seen from the point of view of producing life, seen as another part in a purposeful arrangement of parts, it is one in an extensive series of anthropic coincidences, very finely tuned to yield life.

Similarly, there currently is no plausible lawlike model for the origin of life. In his exploration of the question of the origin of life in *The Fifth Miracle,* the physicist Paul Davies argues that something completely different is needed:

> When I set out to write this book I was convinced that science was close to wrapping up the mystery of life's origin. . . . Having spent a year or two researching the field I am now of the opinion that there remains a huge gulf in our understanding. To be sure, we have a good idea of the where and the when of life's origin, but we are a very long way from comprehending the how.
>
> This gulf in understanding is not merely ignorance about certain technical details, it is a major conceptual lacuna. . . . My personal belief, for what it is worth, is that a fully satisfactory theory of the origin of life demands some radically new ideas.[11]

I suggest that the origin of life is best viewed not as lawlike, but as one more of the long, long chain of anthropic "coincidences" very, *very* finely tuned to yield life. In this view the origin of life was a unique event, like the origin of the moon, and was purposely arranged. For example, just as the origin of the moon involved a particular body of a particular mass traveling at a particular speed and particular angle at a particular time, and so on, so might the origin of life have involved an extensive string of particulars. Perhaps a particular molecule in a primeval ocean hit another at a particular angle when, say, a particular hydroxide ion was close enough to catalyze a particular reaction, and the product of that reaction underwent a long string of other unique, particular events to yield the first cell. Although at the time the molecules may have been following standard physical laws, no law or general conditions were sufficient to cause the origin of life. It was simply a finely tuned, unique event—undoubtedly much, much more finely tuned than the origin of the moon, but another finely tuned event nonetheless.

Although that view might strike some people as strange, if we admit the possibility of an agent who can choose and implement the laws of physics for the universe, then there is no principled reason to think that implementing much greater fine-tuning would be beyond it.

If the origin of life is a finely tuned event, then research directed at uncovering some reproducible set of generic conditions that would yield life will continue to prove futile, as it has for the past half century. Nonetheless, with the abandonment of the fruitless "lawlike" origin-of-life paradigm, science might productively take up an approach similar to that of astronomers studying the unique conditions needed to explain the origin of the moon. The Nobel laureate Francis Crick once remarked, "An honest man, armed with all the knowledge available to us now, could only state that in some sense, the origin of life appears at the moment to be almost a miracle, so many are the conditions which would have had to have been satisfied to get it going."[12] Investigation of those very many, unique, anthropic, fine-tuned conditions needed to start life could keep scientists busy for many years.

Let me be clear. I am not saying the origin of life was simply an extremely improbable accident. I am saying the origin of life was deliberately, purposely arranged, just as the fundamental laws and constants and many other anthropic features of nature were deliberately, purposely arranged. But in what I'll call the "extended-fine-tuning" view, the origin of life is merely an additional planned feature culminating in intelligent life. The origin of life is simply closer to the very same goal that the other, more distant anthropic features (laws, chemical properties, and so forth) were also put in place to bring about. Nonetheless, just as it was possible to discover a set of proximate conditions that would lead to the origin of the moon, it may also be possible to arrange a local set of conditions that would lead to life, and that would be a scientifically interesting project. If it succeeded, some would claim that it revealed that life needed no miracle. But in fact it would show the beginning of life needed a directing intelligence.

DESCENT BY NONRANDOM MUTATION AS
MULTIPLE FINELY TUNED EVENTS

Fine-tuning doesn't stop at life's origin. One can view *all* necessary biological features that are beyond what it's biologically reasonable to expect of unintelligent processes—for example, all cellular protein complexes containing two or more protein-protein binding sites—as just more and more and more examples of the fine-tuning of the universe for life, akin to the unique events that produced the moon. But in these cases the fine-tuning is actually *within* the fabric of life itself. As Darwin thought, life descended with modification from one stage to another. Mutations arose in a long series—but many were not random. After the first DNA was formed by purposeful, anthropic events, felicitous mutation kept piling on felicitous mutation, either one by one or in larger clusters, at just the right times they were needed, in a way we have no statistical right to expect, like cosmic detail after cosmic detail in the universe, like ball after ball in the side pocket, to yield the multiple coherent features we find in the cell.

As with the origin of life, it may be possible for scientists to select proximate physical conditions in the laboratory, and deliberately cause batches of certain mutations to occur at the right times, and that would be a scientifically interesting project. But without the intimate involvement of a directing intelligence, they would not come about in nature.

From what has been learned in the past few decades about the complexity of the genetic basis of animal development, it seems reasonable to think that purposeful design extends into biology at least to the level of the major classes of vertebrates, perhaps further. Figure 10.1 illustrates this view. As the figure suggests, design of the universe at large and design in biology can be viewed as all of a piece—simply the purposeful arrangement of all the surprisingly many parts that science has discovered are necessary for life, both internal and external to it. The planning that went into the laws of the universe or the properties of water is no different in principle from the planning that went into details of the molecular machinery of life. In the extended fine-tuning view I am presenting, general anthropic coincidences are just the very beginning, akin to the smooth pool table in a trick shot. The depth of the black region in Figure 10.1 simply takes into account that scientific progress has shown enormously greater planning is needed for life, far beyond the general laws of nature.

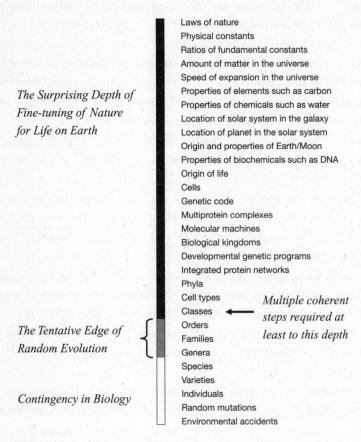

FIGURE 10.1
The surprising degree of fine-tuning of nature required for life on earth.

Deep consilience wells up from a consideration of the basic laws of the universe, the chemistry of certain molecules, terrestrial features, and life itself. That consilience is reminiscent of Isaac Newton's uniting of celestial and terrestrial motion under a single explanation. Just as he demonstrated that the heavens are governed in the same way as the earth, so too has the hard work of many scientists across many scientific disciplines in the past century unexpectedly demonstrated that both the universe at large and the earth in particular were designed for life. The heavens and earth—and life itself—alike are fine-tuned.

That need not have been the case. Compared to modern scientists, nineteenth-century scientists knew little of the depth of fine-tuning in nature. When unknown laws of physics were later discovered, when new properties of elements were elucidated, when the environments of other planets such as Venus and Mars were explored, for all scientists knew or expected, these discoveries might have shown that producing life was not very difficult. When the cell was probed, it might have turned out to be the "simple little lump of albuminous combination of carbon"[13] that Ernst Haeckel innocently expected more than a century ago, rather than the elaborate mechanized factory it is now known to be.

From the far reaches of the universe to the depths of the cell, separate branches of modern science have all discovered astonishing, unexpected fine-tuning—design. As philosopher William Whewell, who coined the term consilience, noted in the nineteenth century, "When an Induction, obtained from one class of facts, coincides with an Induction, obtained from another different class," we can be very confident it is correct.

IS EVERYTHING DESIGNED?

Here's an important question. If design extends from the bottom of physics up to higher levels of biology, is everything in between also designed? Is nothing left to chance? No, there's no scientific reason to think that any but a minuscule fraction of the details of the universe or life are intended. To understand why, let's briefly reconsider the pool table illustration, but with a twist. Suppose on an oversized pool table there were ten numbered red balls and a thousand unnumbered green ones. The cue strikes a ball, which hits another, and at the end of the videotape we see the ten red balls all falling one by one, in numerical order, into the side pocket, but the green balls come to rest scattered on the pool table surface. We could confidently conclude that the trick shot was meant to sink all the red balls, but couldn't say much about the fate of the great majority of green balls. For all we could tell, where the green balls ended up—after playing whatever role they had in affecting the path of the red balls—was of no concern to the pool shark who set up the shot. The path of the red balls to their resting point certainly was designed, but the path of everything else probably was a chance artifact of that plan.

Similarly, although we have compelling evidence that the universe was designed for life, we have no scientific evidence for the design of the details of most inorganic matter. Our nascent world might have benefited from a planned collision in order to prepare it for intelligent life, but there's no reason to think that all—or even many—astronomical collisions in the universe are planned. The very great majority of the universe might proceed on its merry way without any particular relevance to life on earth, even if a prime goal of the universe was to produce intelligent life on earth.

The very great majority of terrestrial biology might proceed the same way, too, without any necessary, direct connection to the goal of intelligent life. The overwhelming number of mutations may be due to chance (to little constructive effect), numerically swamping the comparatively few due to design (which nonetheless are inordinately significant). Explicit design appears to reach into biology to a certain level, to the level of the vertebrate class, but not necessarily further. Randomness accounts perfectly well for many aspects of life. Contingency is real.

LUSH LIFE

Whenever one tries to address the most basic questions, such as where did the universe, life, and mind come from, some of the prospective answers sound strange. Design isn't the only option. There is an alternative response to anthropic arguments: Our universe wasn't designed—we're just lucky. Our universe is just one of many universes, and we happen to live in the right one. Which idea is the stranger of the two?

The Oxford philosopher Nick Bostrom explains the multiple-universe idea this way:

[The multiverse hypothesis, or "ensemble" hypothesis] states that the universe we observe is only a small part of the totality of physical existence. This totality itself need not be fine-tuned. If it is sufficiently big and variegated, so that it was likely to contain as a proper part the sort of fine-tuned universe we observe, then an observation selection effect can be invoked to explain why we see a fine-tuned universe. The usual form of the ensemble hypothesis is that our universe is but one in a vast ensemble of actually existing universes, the totality of which we can call "the multiverse." [14]

In other words, just as medieval astronomers were wrong to take the universe to be only what they could see with their crude telescopes, so might we be wrong to take all of physical reality to be limited to what we can see even with our more advanced instruments. What we have been pleased to call "the universe" might actually be just a tiny part of a much larger "multiverse," which, because it's so big, we can't see.

How might a multiverse help explain fine-tuning? As Bostrom indicates, if a multiverse consisted of a huge collection of relatively isolated universes, each of which was (somehow) randomly assigned values for the laws and constants of nature, the odds might be pretty good that some of the universes would have values that at least allow life. Most philosophers and physicists who think about such things stop there, because they assume life would then be able to get going without too much trouble. As we've seen in this book, that's not the case. Even so, if the number of universes in the multiverse were extraordinarily large, then over that vast space the odds might be pretty good that some of them would have *all* of the extended fine-tuning we find in physics, astronomy, chemistry, and biology. There might be many universes that had the same laws and constants as our own. Most of these would not develop life, but by chance some might have experienced something like a Mars-sized planet hitting a developing earth at the right time. Of those, most would still not develop life, but by extraordinary chance some might have experienced the right molecule collisions for the origin of life. Of those almost all would not develop intelligent life, but again, by extraordinary chance some might have experienced the right mutations at the right times, singly or in large clusters as needed, and intelligent beings would appear.

Would it then be remarkable that we found ourselves in a universe that both permitted and actually developed life? The ensemble hypothesis says no. There might be a zillion dead universes that couldn't develop life, and just one that could and did. But that's the one we simply must be in. By definition we have to exist in a universe that not only was compatible with intelligent life, but also actually developed it, because any universe containing us by definition fits the bill.

Notice that the multiverse scenario doesn't rescue Darwinism. Random mutation in a single universe would still be terribly unlikely as a cause for life. Incoherence and multiple steps would still plague any merely Darwinian scenario in any one universe. In the ensemble hypothesis, the extremely long odds against life are overcome only by brute numbers of universes, not by random mutation and natural selection. Still, although it doesn't help Darwinism, the multiverse scenario would undercut design. If it were true, life wouldn't be due to either Darwin or design. Seen from the proper perspective, it would be one big accident.

Needless to say, the multiverse is speculative.[15] Some physicists have proposed mathematical models that they think might indicate something like a multiverse, but the models are pretty iffy. And some multiverse models themselves require much fine-tuning to make sure that, if real, they would generate universes with the right properties.[16] Nonetheless, let's assume two of the strongest possible general versions of the multiverse scenario and consider some of their serious shortcomings. The two versions: 1) a finite number of random[17] universes in a multiverse; and 2) an infinite number.

Let's assume a multiverse with a tremendously large but finite number of random universes. In some universes life arises by chance, and we, of course, live in a universe that both permits and contains intelligent life. Beyond that, what should we expect of our world, our earth? Statistically, we should predict that the world has taken the fewest possible steps needed to produce intelligent life, and that no life in the world contains any complex, coherent machinery that isn't required, directly or indirectly, to support intelligent life. (This is a game of pure logic, so bear with me.) The reason we should expect it is that the only thing that needs to be special about the universe, by the definition of our theory, is that intelligent life should exist to ask questions and observe what's going on. Beyond that, the universe should be run-of-the-mill.

Here's an analogy to help illustrate. Suppose in a large room were gathered everyone who had won a prize in the past year in the Powerball lottery, no matter how large or small. They were all having a party, closed to the public. You (who haven't won a dime) are an autograph hound and want the signature of a grand prize winner. So you sneak into the guarded room and meet someone at random. What are the odds that the first person you meet is a big-money winner? Very slim.[18] The great majority of folks there will be minimum or small prize winners. That is, a person selected at random from the "winners" category very likely will fulfill just the minimum requirements for getting into the room.

The same goes for universes. On the finite random multiverse view, we should very likely live in a bare-bones world, with little or nothing in life beyond what's absolutely required to produce intelligent observers. So if we find ourselves in a world lavished with extras—with much more than the minimum—we should bet heavily against our world being the result of a finite multiverse scenario. Now let's return from pure logic to the earth as we know it. Is it a bare-bones producer of intelligent life, or is it much more than that? It's difficult to make a rigorous argument on such a question. Yet it certainly seems that life in our world is quite lush and contains much more than what's absolutely needed for intelligence. Just as one familiar example from this book, the bacterial flagellum seems to have little to do with human intelligence, but is tremendously unlikely. If I am correct that it isn't required to produce intelligent observers, then only one in a very large number of universes that had intelligent observers should be expected to also have bacteria with flagella. As a practical matter it's impossible to absolutely rule out that at some point, in the history of life, the flagellum played a crucial role leading to intelligence.[19] But until we find convincing evidence for that, and for the role of much else in biology whose connection to intelligence is obscure, we should regard the finite random multiverse as a beguiling but quite unlikely hypothesis.

BRAIN IN A VAT

What if a multiverse contained not just a tremendous number of universes, but an infinite number? In that case the situation changes utterly—and becomes very weird indeed. Infinity is not just some ultrabig number; it's a completely different and strange case. If the number of universes in the multiverse were infinite, and if all the necessary factors such as laws, constants, and so on could vary in the right ways, there would not just be rare, occasional universes like ours—there would be an infinite number identical to it. There also would be an infinite number of universes almost identical to ours, where everything was the same except for some trivial detail; where, say, instead of hesitating for two seconds after the traffic light turned green last Tuesday morning, you (or your double) hesitated for three seconds. And there would be an infinite number where more and more aspects differed. There would also be an infinite number of dead universes, without the necessary or sufficient conditions for life.

If you think all that's odd, consider this. There would also be an infinite number of universes that did not have what are usually considered the necessary conditions for the gradual development of intelligent life, *but nonetheless contained it.* An infinite number of universes would harbor an infinite number of "freak observers." In his scholarly book *Anthropic Bias* philosopher Nick Bostrom explains:

> Consider a random phenomenon, for example Hawking radiation. When black holes evaporate, they do so in a random manner such that for any given physical object there is a finite (although, typically, astronomically small) probability that it will be emitted by any given black hole in a given time interval. Such things as boots, computers, or ecosystems have some finite probability of popping out from a black hole. The same holds true, of course, for human bodies, or human brains in particular states. Assuming that mental

states supervene on brain states, there is thus a finite probability that a black hole will produce a brain in a state of making any given observation. Some of the observations made by such brains will be illusory, and some will be veridical. For example, some brains produced by black holes will have the illusory of [sic] experience of reading a measurement device that does not exist.

. . . It isn't true that we couldn't have observed a universe that wasn't fine-tuned for life. For even "uninhabitable" universes can contain the odd, spontaneously materialized "freak observer," and if they are big enough or if there are sufficiently many such universes, then it is indeed highly likely that they contain infinitely many freak observers making all possible human observations. It is even logically consistent with all our evidence that *we* are such freak observers.[20]

The Twilight Zone was never so bizarre. In a nutshell: In an infinite multiverse, probabilities don't matter. Any event that isn't strictly impossible will occur an infinite number of times. So (if thinking depends solely on a physical brain), by utter chance in an otherwise dead universe, matter might spontaneously arrange itself into a brain that would contain the true thought, "I am a spontaneously materialized brain in an otherwise dead universe." That will happen a limitless number of times in an infinite multiverse. Matter may also arrange itself into brains with any of an infinite number of false-but-detailed thoughts and memories, such as: "I am a spontaneously materialized brain in a universe that contains one other brain" (but no other brain actually exists there); "I am the Vulcan Mr. Spock with a beard"[21] (but really am a lone brain in outer space); "I am a person reading a book in the twenty-first century United States" (but, again . . .). *All* false thoughts, no matter how detailed, no matter how vivid, will occur without end.

Notice Bostrom's remark, apparently intended without irony: "It is even logically consistent with all our evidence that *we* are such freak observers." But if he is a freak observer, who does he mean by "*we*"? And if you are the freak observer, who is Bostrom? And in either case, what could possibly be meant by "evidence"?

Here is an even-more-outlandish possibility, perfectly consistent with the infinite multiverse scenario: There actually are *no* lawful conditions compatible with the gradual development of intelligent life. There are *zero* possible combinations of laws and constants that allow some orderly progression of a universe that leads to life. The *only* possible "observers" are freaks. An infinite multiverse must contain freak observers, but there's no need for it to contain "real" ones.[22] Don't object that you know of one universe—ours—where such laws and constants do exist. What you are at this very moment "thinking"—as well as any detailed memories you have of the past, no matter how seemingly realistic, including memories of what you may think you know about "nature"—could be due to a random collocation of matter that just popped into existence. The very concepts of "gravity," "protons," "stars"—all you think you know about nature—could be just the pitiful delusion of a freak observer. Reality might be utterly different. Such is the intellectually toxic bequest of the infinite multiverse hypothesis.

There are even stranger possibilities, but there's no need to go further. The point is this: Humans have the extraordinary ability to reason. But in order to take even the first step in reasoning, one must apprehend that reality exists and have confidence that what one perceives with one's senses is generally a reliable reflection of reality. Reality is not something that can somehow be independently verified without begging the paralyzing question. Once reality is doubted, there is no way back. No person—Darwinist, design proponent, or other—who wants to make a rational argument can seriously entertain an idea that pulls the rug out from under reason.

Here's a subtle problem. Some intriguing ideas that at first blush appear reasonable actually contain the poison pill of radical skepticism that undermines reason. One classic notion that undercuts reality is solipsism, which in its pure form holds that nothing exists outside the solipsist's mind—everything else, including other apparent minds, is an illusion, a product of the thinker's own mind. Some other ideas aren't exactly solipsism, but still entail the conclusion that we can know nothing except what's going on in our own minds. A famous example from the history of philosophy is René Descartes's question, How do you know your thoughts aren't being controlled by a demon? A modern version of the same issue might be termed the *Matrix* problem: How do you know you're not really just a brain in a vat, hooked up by scientists to wires that feed you a wholly false perception of reality? The short answer is that you know directly that reality exists. And you must have confidence that your senses are generally reliable. Without those twin, bedrock premises, thought itself is stymied.

Infinite multiverse scenarios are no different from brain-in-a-vat scenarios. If they were true, you would have no reason to trust your reasoning. So anyone who wants to do any kind of productive thinking must summarily reject the infinite multiverse scenario for intelligent life and assume that what we sense generally reflects the reality we know exists. And what we sense, as elaborated through modern science's instruments and our reasoning, is that we live in a universe fine-tuned for intelligent life. Moreover, unlike the lonely solipsist who refuses to recognize the existence of other minds, we can perceive the work of other minds in the purposeful arrangement of parts, which reaches its zenith in the arrangement of the many parts of the universe for life.

Two leading ideas compete to explain fine-tuning in nature: purposeful design or a multiverse. Yet life looks far richer than we have a right to expect on a finite random multiverse scenario, and on an infinite multiverse scenario we have no ability to expect—or even think about—anything. There's every reason to trust our basic human insight that we live in a purposefully designed world.

WHO WAS THAT MASKED MAN?

Whenever we ask fundamental questions such as where the universe came from, how life originated, or what is the nature of mind, we bump into philosophy. Over the remaining sections of the chapter I'll consider several philosophically related topics, starting with the question of who the "agent" might be who designed the universe for life.

Many people are impatient with that question. Since the great majority of the population of the United States and the world believes in God (as a pretty conventional Roman Catholic, so do I), "designer" is often seen as a not-too-subtle code word for God, both by those who like the implications and by those who don't. Although that reaction is understandable among the general public, the leap to God with a capital G short-circuits scholarly arguments that have been going on for millennia across many cultures. Aristotle argued that nature reflected a "Prime Mover," but his conception would scarcely be recognized by adherents of most modern religions. In summarizing the design hypothesis, philosopher Nick Bostrom notes that:

> The "agent" doing the designing need not be a theistic God, although that is of course one archetypal version of the design hypothesis. . . . We can take "purposeful designer" in a very broad sense to refer to any being, principle or mechanism external to our universe responsible for selecting its properties, or responsible for making it in some sense probable that our universe should be fine-tuned for intelligent life.[23]

Like it or not, a raft of important distinctions intervene between a conclusion of design and identification of a designer.

The designer need not necessarily even be a truly "supernatural" being. Consider a peculiar fictional essay published a few years ago in *Nature*, the most prominent science journal in the world, entitled "The Abdication of Pope Mary III . . . or Galileo's Revenge."[24] The gist of the story was that a future pope (yes, a woman) resigns, along with all the cardinals of the Catholic Church, because new scientific evidence proves that the designer of the universe is not a transcendent god. As one character in the tale observes, "The life-generating properties of the very specific fundamental constants that define reality are virtually impossible to explain except as the results of deliberate design." However, "That creator is clearly not the God of the Bible or the Torah or the Qur'an. Rather, the creator is a physicist, and we are one of his or her experiments." Although that scenario may be attuned less to reality than to the amusing fantasies of overgrown science geeks, it does helpfully illustrate that, if one wishes to be academically rigorous, one can't leap directly from design to a transcendent God.

To reach a transcendent God, other, nonscientific arguments have to be made—philosophical and theological arguments. It is not my purpose here to rehearse what has been said over the millennia on that score, or to say why I myself find some of those arguments persuasive and others not. Here I'm content to "take 'purposeful designer' in a very broad sense."

NO INTERFERENCE

How was the design of life accomplished? That's a peculiarly contentious question. Some people (officially including the National Academy of Sciences) are willing to allow that the laws of nature may have been purposely fine-tuned for life by an intelligent agent, but they balk at considering further fine-tuning after the Big Bang because they fret it would require "interference" in the operation of nature. So they permit a designer just one shot, at the beginning—after that, hands off. For example, in *The Plausibility of Life* Marc Kirschner and John Gerhart hopefully quote a passage from an old article on evolution in the 1909 *Catholic Encyclopedia:* "God is the Creator of heaven and earth. If God produced the universe by a single creative act of His will, then its natural development by laws implanted in it by the Creator is to the greater glory of His Divine power and wisdom." [25]

This line of thinking is known as "Theistic Evolution." But its followers are kidding themselves if they think it is compatible with Darwinism. First, to the extent that anyone—either God, Pope Mary's physicist, or "any being . . . external to our universe responsible for selecting its properties"—set nature up in any way to ensure a particular outcome, then, to that extent, although there may be evolution, there is no Darwinism. Darwin's main contribution to science was to posit a mechanism for the unfolding of life that required no input from any intelligence—*random* variation and natural selection. If laws were "implanted" into nature with the express knowledge that they would lead to intelligent life, then even if the results follow by "natural development," nonetheless, intelligent life is not a random result (although randomness may be responsible for other, unintended features of nature). Even if all the pool balls on the table followed natural laws after the cue struck the first ball, the final result of all the balls in the side pocket was not random. It was intended.

Second, "laws," understood as simple rules that describe how matter interacts (such as Newton's law of gravity), cannot do *anything* by themselves. For anything to be done, specific substances must act. If our universe contained no matter, even the most finely tuned laws would be unable to produce life, because there would be nothing to follow the laws. Matter has unique characteristics, such as how much, where it is, and how it's moving. In the absence of specific arrangements of matter, general laws account for little.

Finally, a particular, complex outcome cannot be ensured without a high degree of specification. At the risk of overusing the analogy, one can't ensure that all the pool balls will end up in the side pocket just by specifying simple laws of physics, or even simple laws plus, say, the size of the pool table. Using the same simple laws, almost all arrangements of balls and almost all cue shots would not lead to the intended result. Much more has to be set. And to ensure a livable planet that actually harbors life, much more has to be specified than just the bare laws of physics.

Some people who accept design arguments for physics, but not for biology, nurture an aesthetic preference that our universe *should* be self-contained, with no exceptions to physical laws after its inception. The prospect of the active, continuing involvement of a designer rubs them the wrong way. They picture something like a big hand flinging a Mars-sized orb at the nascent earth, or pushing molecules around, and it offends their sensibilities. Some religious people, in particular, are repelled by that view, thinking it somehow undignified.

Well, we all have our own aesthetic preferences. It's the job of science, however, to try to determine what type of universe actually exists. Like it or not, the more science has discovered about the universe, the more deeply fine-tuning is seen to extend—well beyond laws, past details, and into the very fabric of life, perhaps beyond the level of vertebrate classes. If that level of design required continuing "interference," that's what it required, and we should be happy to benefit from it.

But the assumption that design unavoidably requires "interference" rests mostly on a lack of imagination. There's no reason that the extended fine-tuning view I am presenting here necessarily requires active meddling with nature any more than the fine-tuning of theistic evolution does. One can think the universe is finely tuned to *any* degree and still conceive that "the universe [originated] by a single creative act" and underwent "its natural development by laws implanted in it." One simply has to envision that the agent who caused the universe was able to specify from the start not only laws, but much more.

Here's a cartoon example to help illustrate the point. Suppose the laboratory of Pope Mary's physicist is next to a huge warehouse in which is stored a colossal number of little shiny spheres. Each sphere encloses the complete history of a separate, self-contained, possible universe, waiting to be activated. (In other words, the warehouse can be considered a vast multiverse of possible universes, but none of them have yet been made real.) One enormous section of the warehouse contains all the universes that, if activated, would fail to produce life. They would develop into universes consisting of just one big black hole, universes without stars, universes without atoms, or other abysmal failures. In a small wing of the huge warehouse are stored possible universes that have the right general laws and constants of nature for life. Almost all of them, however, fall into the category of "close, but no cigar." For example, in one possible universe the Mars-sized body would hit the nascent earth at the wrong angle and life would never commence. In one small room of the small wing are those universes that would develop life. Almost all of them, however, would not develop intelligent life. In one small closet of the small room of the small wing are placed possible universes that *would* actually develop intelligent life.

One afternoon the überphysicist walks from his lab to the warehouse, passes by the huge collection of possible dead universes, strolls into the small wing, over to the small room, opens the small closet, and selects one of the extremely rare universes that is set up to lead to intelligent life. Then he "adds water" to activate it. In that case the now-active universe is fine-tuned to the very great degree of detail required, yet it is activated in a "single creative act." All that's required for the example to work is that *some* possible universe could follow the intended path without further prodding, and that the überphysicist select it. After that first decisive moment the carefully chosen universe undergoes "natural development by laws implanted in it." In that universe, life evolves by common descent and a long series of mutations, but many aren't random. There are myriad Powerball-winning events, but they aren't due to chance. They were foreseen, and chosen from all the possible universes.

Certainly that implies impressive power in the überphysicist. But a being who can fine-tune the laws and constants of nature is immensely powerful. If the universe is purposely set up to produce intelligent life, I see no principled distinction between fine-tuning only its physics or, if necessary, fine-tuning whatever else is required. In either case the designer took all necessary steps to ensure life.

Those who worry about "interference" should relax. The purposeful design of life to any degree is easily compatible with the idea that, after its initiation, the universe unfolded exclusively by the intended playing out of natural laws. The purposeful design of life is also fully compatible with the idea of universal common descent, one important facet of Darwin's theory. What the purposeful design of life is *not* compatible with, however, is Darwin's proposed mechanism of evolution—*random* variation and natural selection—which sought to explain the development of life explicitly without recourse to guidance or planning by anyone or anything at any time.

BY ANY OTHER NAME

Is the conclusion that the universe was designed—and that the design extends deeply into life—science, philosophy, religion, or what? In a sense it hardly matters. By far the most important question is not what category we place it in, but whether a conclusion is true. A true philosophical or religious conclusion is no less true than a true scientific one. Although universities might divide their faculty and courses into academic categories, reality is not obliged to respect such boundaries. Understanding some aspects of the real world might require multiple modes of reasoning. Still, in what category is it best to pigeonhole design?

First, is any conclusion of design necessarily religious? Pope Mary didn't think so, and she's infallible.

Then is design a philosophical conclusion, a scientific one, or maybe both? Here's where serious opinions diverge. On the one hand, some people think that, if a conclusion implicates an intelligent agent (even merely a human mind), or if it threatens to point beyond nature, it's better classified as philosophy. On the other hand, I regard design as a completely scientific conclusion. For many years philosophers have struggled to come up with an airtight definition of science, without much luck. But as a rough-and-ready definition, I count as "scientific" any conclusion that relies heavily and exclusively on detailed physical evidence, plus standard logic. No relying on holy books or prophetic dreams. Just the data about nature that is publicly available in journals and books, plus standard modes of reasoning.

If that's one's definition of a scientific conclusion, then design fits to a tee. The public data for design come from many branches of science—physics, astronomy, biology. The reasoning that leads to a conclusion of design for the universe and life is the same kind of reasoning that leads to a conclusion of design for anything—perceiving a purposeful arrangement of parts. The strength of the reasoning is publicly acknowledged, at least in regard to the general laws of the universe, by many scientists and philosophers.

PREDICTION AND TESTING

If my minimalist definition is the rough standard, then design is certainly science. However, some opponents of design demand two additional qualities of a scientific idea that they believe disqualify it. First, they say, design theory isn't testable. It does not make specific predictions. Second, say the critics, design theory states that certain events happened by mysterious means that we cannot explain. They object to a videotape that never spots the pool shark.

Coming from Darwinists, both objections are instances of the pot calling the kettle black. Darwinism does not have a consistent record of confirmed predictions; quite the opposite. An eminent leader of the neo-Darwinian synthesis declared forthrightly a half century ago that "the search for homologous genes is quite futile." Later work showed that "the view was entirely incorrect." The president of the National Academy of Sciences stated that "the chemistry that makes life possible is much more elaborate and sophisticated than anything we students had ever considered." And since the new field of evolutionary developmental biology has led to big surprises—"The most stunning discovery of Evo Devo . . . was entirely unanticipated"—we're justified in thinking that the theory that guided all these expectations was wrong. Yet what price has Darwinism paid for misleading scientists? Those who overlook the falsified predictions of their own theory are in no position to demand hard-and-fast predictions of another.

The same goes for the call to, essentially, produce a high-quality videotape of the pool shark in action. That demand is often issued by the same people who excuse themselves from identifying the detailed steps that random mutation and natural selection putatively would take to complex biological structures. Reviewing *Darwin's Black Box* in 1996 for *Nature*, University of Chicago evolutionary biologist Jerry Coyne wrote, "There is no doubt that the pathways described by Behe are dauntingly complex, and their evolution will be hard to unravel. . . . We may forever be unable to envisage the first proto-pathways." If anyone thought it was hard to unravel ten years ago, it's far worse now. Those who stick with Darwinism even if they can't rigorously envisage supposed random pathways to complex systems are in no position to demand that design theorists escort the designer to the next science conference.

Both additional demands—for hard-and-fast predictions or for direct evidence of a theory's fundamental principle—are disingenuous. Philosophers have long known that no simple criterion, including prediction, automatically qualifies or disqualifies something as science, and fundamental entities invoked by a theory can remain mysterious for centuries, or indefinitely. Isaac Newton famously refused to speculate about the nature of gravity; the basis for biological variation remained hidden for a century after Darwin; the cause of the Big Bang remains unknown even today. The strident demands heaped on the head of intelligent design by those hostile to it are simply attempts at verbal gerrymandering—trying to win by words what can't be won by evidence. Yet science is not a word game that's decided by definitions—it's an unsentimental, no-holds-barred struggle to understand nature.

Of course, although prediction and testing aren't nearly as straightforward as some simplistically assert, they nonetheless do play a critical role in science. If a theory has no implications for nature at all, or is completely disconnected from testing, then one can never have confidence that it is correct. Although philosophers of science agree that it is virtually impossible to falsify a theory directly, some tests of any theory's basic expectations can make it far less credible. The Michelson-Morley experiment didn't directly falsify the theory of the ether. The theory still might have been tinkered with, so that the failed experiment could somehow be shown to be consistent with it. But the failure to find an expected major effect of the ether severely and rightly shook scientists' confidence.

The intensive studies of malaria discussed in this book are the equivalent of a Michelson-Morley experiment for Darwinism. Darwinism implicitly entails the strong, broad, basic claim that, given enough chances, random mutation and natural selection can build the sorts of complex machinery we see in the cell. Intelligent design implicitly entails an equally strong, broad, basic prediction, that random mutation *cannot* do so. Design denies not only that *some* specific piece of machinery (say, the bacterial flagellum) would be produced by random mutation, but that *any* complex, coherent molecular machinery would. Although random processes can account for small changes, there are real limits. Beyond those limits, design is required.

Darwin and design hold opposite, firm expectations of what we should find when we examine a truly astronomical—a hundred billion billion—number of organisms. Up until recently, the magnitude of the problem precluded a definitive test. But now the results are in. Darwinism's most basic prediction is falsified.

THE EDGE OF PUBLIC HEALTH

Squabbles about what makes a theory scientific interest mainly philosophers. Does design make any practical difference? If it doesn't, then why should anyone care?

The question is misbegotten. Although some people value science chiefly for the control it affords us over nature or the technological benefits it brings, that's not its primary mission. The purpose of science is simply to understand the universe we live in, for its own sake. If that understanding leads to practical benefits, great. If not, that's okay, too. Science is an intellectual adventure, not a business trip. If at the end of the scientific day we simply know more about the world than at the beginning, our chief goal has been met.

Nonetheless, although a scientific theory doesn't have to have important practical implications, intelligent design does have them. As we've seen, nature plays hardball. A million people a year, mainly small children, die from malaria. Many more die from HIV and other infections. In order to counter such biological threats, we have to use every scrap of knowledge we have. We must understand *both* the capabilities *and* the limitations of nature.

In recent years, to educate the public about the medical importance of Darwin's theory, some scientific organizations have emphasized the role of random mutation and natural selection in the development of antibiotic resistance. They are quite right to do so. Tiny, single changes in a target protein can destroy its ability to bind an antibiotic, rendering the antibiotic ineffective. For public health purposes, that's a critical biological fact to understand.

But antibiotics that require multiple changes are far more resistant to Darwinian processes. *That's a critical fact to understand, too.* Malaria requires several mutations to deal with chloroquine, so it's a far better drug than ones that are stymied by a single mutation. And chloroquine is not the only case. Recently, former University of Rochester microbiologist Barry Hall examined various antibiotics in a class called "carbapenems," which are chemically similar to penicillin.[26] With unusual clarity of thought on the topic of evolution, Hall wrote, "*Instead of assuming* that [the chief kind of enzyme that might destroy these antibiotics] will evolve rapidly, it would be highly desirable to accurately predict their evolution in response to carbapenem selection" (emphasis added). Using clever lab techniques he invented, he showed that, although most of the antibiotics quickly failed, one didn't. The reason is that neither single nor double point mutations to the enzyme allowed it to destroy the certain antibiotic (called "imipenem"). Wrote Hall, "The results predict, with >99.9% confidence, that even under intense selection the [enzyme] will not evolve to confer increased resistance to imipenem." In other words, more than two evolutionary steps would have to be skipped to achieve resistance, effectively ruling out Darwinian evolution.

If antibiotics could be found that required a double CCC to counter, they would likely *never* lose their effectiveness.

On matters of public health, Darwin counsels despair. A consistent Darwinist must think that random mutation will get around *any* antibiotic eventually—after all, look at all that magnificent molecular machinery it built. . . . But intelligent design says there's always real hope. If we can find the right monkeywrench, just one degree more difficult to oppose than chloroquine, it could be a showstopper.

In dealing with an often-menacing nature, we can't afford the luxury of elevating anybody's dogmas over data. In medical matters, it's critical that we understand what random mutation can do. And it's equally critical that we locate the edge of evolution.

WHEN BAD THINGS HAPPEN TO GOOD PEOPLE

Here's something to ponder long and hard: Malaria was intentionally designed. The molecular machinery with which the parasite invades red blood cells is an exquisitely purposeful arrangement of parts. C-Eve's children died in her arms partly because an intelligent agent deliberately made malaria, or at least something very similar to it.

What sort of designer is that? What sort of "fine-tuning" leads to untold human misery? To countless mothers mourning countless children? Did a hateful, malign being make intelligent life in order to torture it? One who relishes cries of pain?

Maybe. Maybe not. A torrent of pain indisputably swirls through

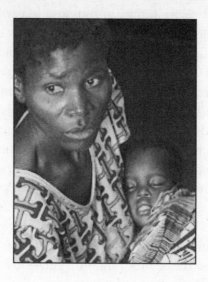

FIGURE 10.2
A Malawian mother holds her malarious child. (Courtesy of Stephenie Hollyman,
www.blazingcontent.com.)

the world—not only the world of humans but the world of sentient
animal life as well. Yet, just as undeniably, much that is good graces
nature. Many children die, yet many others thrive. Some people languish, but others savor full lives. Does one outweigh the other? If
so, which outweighs which? Or are pleasure and pain, good and
evil, incommensurable? Are viruses and parasites part of some brilliant, as-yet-unappreciated economy of nature, or do they reflect the
bungling of an incompetent, fallible designer?

Whether on balance one thinks life was a worthwhile project or
not—whether the designer of life was a dope, a demon, or a deity—
that's a topic on which opinions over the millennia have differed
considerably. Each argument has some merit. Of the many possible
opinions, only one is really indefensible, the one held by Darwin. In
a letter to Asa Gray, he wrote: "I cannot persuade myself that a
beneficent and omnipotent God would have designedly created the
Ichneumonidae with the express intention of their feeding within
the living body of caterpillars."

Wasp larvae feeding on paralyzed caterpillars is certainly a disquieting image, to say nothing of malaria feeding on children. So did Darwin conclude that the designer was not beneficent? Maybe not omnipotent? No. He decided—based on *squeamishness*—that no designer existed. Because it is horrific, it was not designed—a better example of the fallacy of *non sequitur* would be hard to find. Revulsion is not a scientific argument.

Darwin could have learned something from the hard-boiled Yiddish proverb, "If God lived on earth, people would break his windows." Maybe the designer *isn't* all that beneficent or omnipotent. Science can't answer questions like that. But denying design simply because it can cause terrible pain is a failure of nerve, a failure to look the universe fully in the face.

THE TRUMAN SHOW

In the late 1990s the actor Jim Carrey starred in a clever movie called *The Truman Show*. Truman Burbank is a man who was raised since birth in a city on an island he has never left. At the start of the movie he's a thirtyish, married, childless insurance salesman leading an unremarkable life. He goes to work, listens to the radio in his car, is nagged by his mom about grandchildren, and endures the usual quotidian joys and sorrows. Unbeknownst to himself, however, Truman is actually the star of TV's biggest reality show. The entire island is a set—concealing thousands of miniature TV cameras that broadcast his every move to a faithful audience—and the island's population, including his wife and best friend, are actors. What Truman at first innocently takes as an unplanned world is actually elaborately designed around him. As the movie unfolds Truman becomes increasingly suspicious, finally figures it all out, and leaves the island.

So, is our universe *The Truman Show* writ large? Is humanity playing Jim Carrey's role, and is the earth the island? If so, what might that mean for our quotidian lives? Are we really just puppets, pulled this way and that by invisible strings?

From what science has discovered, the universe is indeed elaborately designed around us, so in that sense the earth really is much like Truman's island, strange as that may seem. But even so, from the bare conclusion of design, I see no necessary major implications for our daily lives. Even on Truman's artificial island, he made up his own mind, overcame or yielded to his fears, decided when to stay or go. When he became suspicious that the island was designed, he didn't flinch; he strove to discover the facts. In the most basic sense, within the borders of the set, he lived his own life.

And we live our own lives. We have as much control over our daily lives as did people in the nineteenth century, before the fine-tuning of nature was discovered. Within the boundaries of the society in which we participate—family, friends, culture—we make our own decisions, and enjoy or suffer the consequences, as we always have. Unlike on the movie island, our neighbors are not actors. They are other striving people, and our choices can affect not only ourselves, but them, too. In that regard, at least, the progress of science has changed nothing of our daily lives. We have as much opportunity to do right or wrong, to despair or hope, to help or hurt, as we ever did.

All the world may indeed be a stage, as Shakespeare wrote. Without a stage there would be no play and no actors. Yet the stage seems set for improvisational theater. The actors' lines are spontaneous, not scripted, and, on that dangerous, living stage, they make of the play what they will.

I, Nanobot

Understanding the immense hurdles facing random mutation requires at least a passing familiarity with aspects of the molecular foundation of life. For those who might not remember their high school biology, in this appendix I present—as minimally, painlessly, and entertainingly as I can—a thumbnail sketch of the structure of protein and DNA, and an outline of how they work.

GRAY GOO

An electronic computer in the 1940s would fill a large room. By the late 1970s the antiquated clunkers had given way to personal computers that were thousands of times faster and could fit comfortably on a desktop. Today's computers are thousands of times more powerful than the first PCs, and even smaller and sleeker. Faster, smaller, better—the trend for electronic machines seems relentless. Where will it lead? Some futurists have envisioned a, well, future when humanity can construct machines on the atomic scale. Molecular-sized robots will manipulate molecules, the idea goes, to build infinitesimally small machines that can themselves manufacture other machines.[1] The field of the tiniest machines has been dubbed "nanotechnology."

Tiny robots might do humanity much good. Yet in his 1986 book *Engines of Creation* Eric Drexler worried about the dark side: What if self-replicating nano-sized robots (nanobots) escaped the lab? The nanobots might replicate uncontrolledly, eating everything in sight, becoming an ever-expanding "gray goo" that takes over the universe. With admirable understatement he warned, "We cannot afford certain kinds of accidents."

Today the field of nanotechnology is hot, but as reported in *Nature,* the sci-fi worries of Drexler from twenty years ago still dog workers (stirred up by the 2002 Michael Crichton novel *Prey* that detailed a gooey catastrophe).[2] "Nanotechnology researchers are sick of hearing about 'grey goo.' Their research is still largely speculative, yet the notion that swarms of tiny self-replicating robots could escape from laboratories and destroy our world comes up time and time again when nanotechnology is discussed with the public." Drexler himself, weary of the hysteria, recently avowed, "I wish I had never used the term 'grey goo.'"

But just imagine—self-replicating nano-scale robots! Robots that can manipulate single molecules at a time! Tiny robots that could fill the earth! Wow, what a glorious future it will be—a glorious future that looks a lot like the glorious present and the glorious past, where nanobots already do all those things, and have been doing them for billions of years. You see, in biology nanobots are called "cells."

Most people don't think of cells as robots, probably because cells are made of organic materials rather than metal. But cells truly are self-replicating nanoscale robots.[3] Self-replicating, because of course they reproduce themselves. Nanoscale, because most cells are quite tiny and all can manipulate single molecules. Robots, because their activities are carried out unconsciously and automatically by precision machinery that follows ordinary physical laws. And like Drexler's frightful gray goo, biological nanobots would be more than happy to take over the world. Consider that a few single-celled malarial parasites injected into a human by the bite of a mosquito can multiply to a trillion in a short time, consuming much of the victim's blood in the process. They would gleefully fill the earth if they could.

For most of our history humanity did not comprehend that the earth was filled with nanobots, or that assemblies of nanobots composed the mysterious creatures that could be seen by the unaided eye—mushrooms, lobsters, turnips, catfish, people. The realization that gray goo (or rainbow goo, anyway) had already taken over the world dawned on us slowly, after centuries of investigation. To drive home the critical point that the foundation of life is a congeries of ultrasophisticated molecular machinery gathered inside the nanobots called cells—and to give some background for showing what Darwinian evolution can and can't do in the realm of the nanobots—in this appendix I'll recount some highlights from the history of biology and take a look at how some work gets done in a real nanobot. We'll start with the large and work downward because, in one sense, the science of biology did what computer science has done more recently: focused first on big machinery and then worked its way down to the nano-scale.

THE PROMISE AND THE PERIL

Most people these days can learn of the basic underpinnings of life in a year, usually in their high school biology class. But it was not always so. It is only because of centuries of work by dedicated naturalists that we can open a book and learn about the different categories of plants and animals, the organization of the circulatory system, the structure of the vertebrate eye, the genetic code, the action of muscles, the chemical basis of life, and so on. Thousands of years ago no biology textbooks had been written, and the ancients had to puzzle out the structure of life for themselves.[4] The first firm steps on that long, hard, sometimes dangerous path arguably were taken by the Greek philosopher Aristotle. Aristotle knew that to understand nature, you had to pay close attention to it. And very close attention he did pay. Consider the formidable powers of observation reflected in his description of octopus reproduction.[5]

> The octopus breeds in spring, lying hid for about two months. The female, after laying her eggs, broods over them. She thus gets out of condition since she does not go in quest of food during this time. The eggs are discharged into a hole and are so numerous that they would fill a vessel much larger than the animal's body. After about fifty days the eggs burst. The little creatures creep out, and are like little spiders, in great numbers. The characteristic form of their

limbs are not yet visible in detail, but their general outline is clear. They are so small and helpless that the greater number perish. They have been so extremely minute as to be completely without organization, but nevertheless when touched they move.

This passage illustrates both the promise and the peril of simple observation. The promise is that, by watching attentively, one can learn much about life. The peril is that, even if you do look as closely as you can, not everything of importance is visible. The philosopher thought that baby octopuses, because they are so small, are "completely without organization." Wrong! Nanobots are nothing if not organized. But the organization of nanobots cannot be seen with the naked eye. To unaided vision the intricate but minute machinery looks just like gray goo.

Although a measure of progress was made by Aristotle and other ancient naturalists, their inability to see down to tiny scales often led them astray. Perhaps the most spectacular blunder was committed by the Roman physician Galen.[6] Galen knew that the heart was a pump, but what happened to the blood that the heart pumped? Unable to see that large arteries lead to tiny arterioles that lead to microscopic capillaries that lead to minute venules that lead to visible veins and then back to the heart in a closed loop, Galen could only guess that the blood pumped out of the heart drained into the tissues to "irrigate" them. His mistaken idea was taught to medical students for more than a thousand years.

Technical innovations were needed to overcome the limitations of human eyesight, to make the details of life visible. The first major breakthrough was the microscope, initially put to consistent scientific use in the seventeenth century. Based on theoretical considerations,[7] blood circulation had first been hypothesized by William Harvey in 1628—the year Marcello Malpighi was born. Among many other discoveries, through a microscope the grown-up Malpighi observed otherwise-invisible capillaries that connected the larger blood vessels. So a technical advance—the microscope—proved that blood circulated and corrected Galen's thousand-year-old mistake.

As important as was the discovery of the circulation of blood, however, the overarching significance of the microscope lay in its unveiling of a completely unsuspected, invisible level of life—the micro level. Aristotle thought that baby octopuses were formless, yet microscopes revealed their intricate form. Insects were thought to lack internal organs, but microscopes showed them aplenty. With the ability to see more and more detail, a clearer understanding of life was emerging. Sometimes, however, even though they could be seen, the importance of microscopic details remained obscure. Some of the earliest seventeenth-century microscopic work showed that plant tissues were built of little units with distinct borders—cells.[8] It wasn't until the mid nineteenth century, however, that the German scientists Matthias Schleiden and Theodor Schwann hypothesized that the cell was actually the basic unit of life, that it was in some sense an independent system, and that all living things were composed of cells and their secretions. Bingo!

Science had glimpsed the cellular nanobot through a microscope, but still was far from comprehending it. The reason for befuddlement was that, although microscopes can image objects a bit smaller than cells, even microscopes can't make visible in sufficient detail the molecular machinery of the cell, whose components are very much smaller than the cell itself. To the microscopes of the nineteenth century the cell looked like "a simple little lump of albuminous combination of carbon."[9] In other words, like gray goo. To allow us to understand the complex workings of the cell, techniques had to be developed that could press beyond the micro scale down to the nano scale. That took another hundred years, until the middle of the twentieth century.

Shortly after World War II a new technique allowed science to peer directly into the nanomachinery of the cell. X-ray crystallography involves shining a focused beam of X-rays onto a crystal of a pure molecular substance. The short-wavelength light interacts with the regularly repeating molecules in the crystal in such a way that the diffraction pattern can reveal the exact atomic structure of the repeating molecule. The procedure is always technically challenging. But for molecules containing many thousands of atoms, as molecules from the cell usually do, crystallography at midcentury was horrendously difficult. Nonetheless, after decades of determined effort, in 1959 a small band of scientists correctly deduced the precise structure of one of the simplest molecular machines of the cell—a molecule called myoglobin.

STONE UGLY

"Could the search for ultimate truth really have revealed so hideous and visceral-looking an object?" lamented the Nobel laureate biochemist Max Perutz when he first beheld the irregular, bowel-like structure of myoglobin.[11] Yet, like the mechanical innards of a robot, myoglobin is built to do a job, not to look pretty. Myoglobin belongs to the class of biological molecules called proteins.[12] With a few exceptions, the machinery of the cell consists of assemblies of proteins or, less frequently, individual proteins. Proteins are quite literally the gears and levers, wires and circuits of the nanobot.

In order to understand what natural selection may or may not be able to do with life, we need to familiarize ourselves with the fascinating machinery of the cell: proteins—what they're made of, what they look like, and the ways by which they carry out the vital tasks of the cell. Over the next few sections I'll touch on how proteins work.[13] Don't worry about remembering technical details. There is no exam at the end of the chapter. The point here is just to show you that, like bigger machines, proteins work by mechanical principles.

Although most people think of them just as something you eat— one of the major food groups—when they aren't being eaten proteins are the machinery of the cell, the tools that allow the cell to perform the work of life. Like a nano–Home Depot, human cells contain thousands of different kinds of protein tools. One example of a protein is collagen, a major component of connective tissue. Three collagen molecules intertwine to form a ropelike structure, which is the basis for much of the mechanical strength of skin. Another protein is rhodopsin, which is found in cells that make up the retina of the eye. Rhodopsin's job is to capture photons of light in the initial events of vision. A protein called Ras acts as a switch that helps the cell decide whether it's time to divide or not. When Ras gets damaged, sometimes cancer can develop. Glutamine synthetase is a member of a class of proteins called enzymes, which are chemical catalysts that build and break down the many different chemical compounds the cell requires. As you can see even from this short list, proteins perform an amazing variety of tasks in a cell. However, just as a sewing machine can't be used as a food processor and vice versa, collagen can't be used in vision, and rhodopsin can't strengthen skin. Just like the tools at Home Depot, a given protein is only good for a certain, narrow task.

All proteins are chains that are constructed from a set of just twenty different kinds of small molecules called amino acids (the "building blocks" of proteins) linked together. The difference between two proteins is just the difference between the number and arrangement of the links in the chain—the different kinds of amino acids making them up. A good analogy is to the alphabet and words. (In fact, in scientific communications amino acids are often abbreviated as single letters.) The English alphabet has just 26 letters, but the letters can be put together in a very large number of ways to generate many different words. For example, the word "goo" is made of just three letters. The word "antidisestablishmentarianism" is made of 28 letters. A typical protein "word" has anywhere from fifty to a thousand amino acid "letters" in it. For example, human myoglobin has 153 amino acids while albumin has 609. The first five amino acids in human myoglobin are G-L-S-D-G, while the first five in albumin are D-A-H-K-S.

Where does DNA fit into this picture? DNA carries the information that tells the cell how to build each and every protein it contains. Like proteins, DNA is a linear chain of a limited number of "building blocks," but in the case of DNA there are only four kinds of building blocks (called "nucleotides"). The sequence of nucleotides in DNA directly determines ("codes for") the sequence of amino acids in a protein. Generally a DNA chain in the cell is very long—much, much longer than protein chains.[14] The long DNA chain contains many discrete regions, called genes, each of which codes for a different protein. So one DNA chain can code for many protein chains; in other words, one DNA chain contains many genes. In order for a protein to mutate—that is, in order for the protein to have an altered amino acid sequence—the DNA coding for that particular protein has to change. Mutations, therefore, are fundamentally changes in the DNA sequence coding for a protein; the change in the DNA then causes the cell to produce a changed protein. Here's an analogy. DNA is like a set of instructions to build a machine; a protein is the machine. If the instructions are altered, then an altered machine is produced.

Analogies only take you so far. Although we often rightly speak of the power of words, proteins have abilities that words lack. Unlike words, proteins are physically active—they have palpable powers that can affect their environment. The physical prowess of a protein results from two features: the chemical properties of the twenty different kinds of amino acids it contains, and the exact three-dimensional arrangement of the amino acids of the protein. We should pay special attention to the latter feature—a protein's 3D shape. Just as in our everyday world the shape of a machine part critically affects its ability to perform its job, so, too, for protein machines. Metal forged into a gear of the right size can help a clock to work; a shapeless blob of metal can't. A chain of amino acids—a protein—that folds into the right shape can be part of a molecular machine; with the wrong shape it can't. But what makes a protein fold into the correct shape?

The twenty amino acids can be categorized into several different groups. Some amino acids are oily ("hydrophobic"—water-fearing) and tend to try to avoid water, while others are like sugar and prefer to be dissolved in close contact with water ("hydrophilic"—water-loving). Some amino acids are negatively charged while others are positively charged. These different chemical properties cause different regions of a chain of amino acids—a protein—to attract or repel each other, somewhat like the north and south poles of many tiny magnets. The oily parts huddle together to shield each other from water, water-loving groups strive to stay in touch with water, negative charges attract positive charges, and like charges push away like charges.

The chains of amino acids found in cells—that is, natural proteins—are quite special. If you just randomly linked amino acids into a chain, the result of all these different forces—the pushing and pulling—would very likely be a mess.[15] That is, no particular shape or properties would likely result. In order to form a precisely defined shape that allows a given protein to do a cellular task, the amino acids in biological proteins are arranged so that the attractive and repulsive forces bring together parts of the protein chain that need to be together and push apart regions that need to be apart. Rather than a floppy chain, a protein in the cell folds itself into a compact, active shape. An analogy might be made to a chain of differently shaped magnetic blocks that automatically folds itself into a correctly solved Rubik's cube, and in doing so gains the power to do something special (say, to fit into a larger, more intricate puzzle). If something goes wrong with the folding process, if by accident the protein does not achieve the shape that it's supposed to, then usually it loses all of its special activity. A melted gear can no longer help a clock to tick. A misfolded protein chain has no more power than do, say, the proteins of a fried egg, which can no longer help build a chick. It is the exact shape of each kind of protein, plus the chemical features of their amino acids, that allow proteins to do the marvels they do.

MYOGLOBIN UP CLOSE

To illustrate how one protein works, let's look up close and personal at myoglobin, the first protein whose exact structure was determined. Myoglobin binds oxygen and stores it in muscles; it's especially abundant in the muscles of diving animals such as whales that have to endure long times between breaths. The protein chain of human myoglobin has 153 amino acids, 22 of which are positively charged, 22 negatively charged, 32 water-loving, and 57 water-fearing.[16] In eight segments of the protein chain, the amino acids are arranged so that roughly several oily ones are followed by a few water-loving ones, which are followed by several more oily ones, and so on. This arrangement allows the segment to wrap into a spiral in which one side of the helix has mostly oily amino acids and the other side mostly water-loving ones. The helical segments are stiff but the portions of the chain between the helical segments are rather flexible, allowing the helical segments to fold toward each other. Happily, separate segments can now interact and press their oily sides against each other in the interior of the now compactly folded protein, shielding them from water. Their water-loving, hydrophilic sides face outward to contact water. When all is said and done, the myoglobin chain has folded itself into the exquisitely precise form shown in Figure A.1.

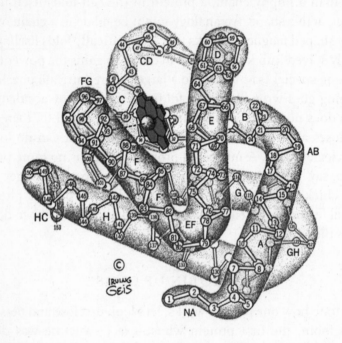

FIGURE A.1

A drawing of myoglobin by the late scientific illustrator Irving Geis. The numbered balls (encased in gray shading) connected by rods are the amino acid positions of the protein. (For clarity, details of the structure of the amino acids are not shown.) The flat structure in the middle is the heme. The sphere in the center of the heme is an iron atom. The letters mark different helices and turns in the protein. The folded shape of the protein is required for it to work. (Illustration by Irving Geis. Rights owned by Howard Hughes Medical Institute. Reproduced by permission.)

The shape of the folded myoglobin allows it to bind tightly to a small, rather flat molecule with a hole in its center. The molecule is called "heme" (let's not worry about where heme comes from). The heme itself is rather oily and fits into an oily pocket formed by the folded myoglobin, like a hand fits into a glove. Now, the heme is also the right size, and has the right chemical groups, to tightly bind one iron atom in its central hole. When the heme fits into the myoglobin pocket, a particular amino acid (the histidine at the eighty-seventh position in the protein chain; histidine is abbreviated as "H") from the myoglobin is precisely positioned to hook onto the iron and keep the heme in place. The iron in heme can bind ("coordinate") to six atoms. Four of those atoms are provided by the heme itself, and one is from the myoglobin's "H". That leaves one position of the iron open to bind another atom. The open position can tightly bind oxygen when it's available. All those features combine to allow myoglobin to fulfill its assumed [17] role as an oxygen-storage protein in muscle tissue.

Again, don't worry about remembering those technical details. As far as this book is concerned, the most important point for us to notice here is that myoglobin does its job entirely through mechanistic forces—through positive charges attracting negative ones, by a pocket in the protein being exactly the right size for the heme to bind, by positioning groups such as "H" in the very place they are needed to do their jobs. Proteins such as myoglobin don't work through mysterious or novel forces, as they once were thought to do. They work through well-understood ones, like the forces by which machines in our everyday world work, like the forces that will control artificial nanobots, should they ever be built. A crucial conclusion is this: Because biological molecular machines work through forces we understand pretty well, we can judge pretty well which improvements are likely to be able to be made to them by random mutation and natural selection, and which are likely to be unattainable.

BEYOND MYOGLOBIN

Believe it or not, myoglobin is one of the smallest, simplest proteins of the nanobot. What's more, myoglobin works alone, which is unusual among proteins. Most proteins work in teams where each protein fits together with others in a sort of super Rubik's cube, and each has its own role to play in the team's task, much as a particular wire or gear might have its own role to play in, say, a time-keeping mechanism in a robot. To give a taste of such teamwork, in this section I'll briefly discuss the workings of a protein system that is related to, but somewhat more complicated than, myoglobin.

Myoglobin stores oxygen in muscle, but a different protein, called hemoglobin, transports oxygen in red blood cells from the lungs to the peripheral tissues of the body. Although in many ways it is similar to myoglobin, hemoglobin is more complex and sophisticated. Hemoglobin is a composite of four separate protein chains, each one of which is approximately the same size and shape as myoglobin, each one of which has a heme group that can bind an oxygen molecule as myoglobin does. So hemoglobin is about four times larger than myoglobin. The four chains of hemoglobin consist of two pairs of identical chains: two "alpha" chains and two "beta" chains. (Here's a point of terminology: When several chains of amino acids come together to do a job, and if they generally stay together for the lifetime of the cell, the whole complex of the several chains is referred to as the protein, and each of the chains alone is called a "subunit.") The sequence of amino acids in both the alpha and beta subunits is similar to, but not identical with, the sequence of amino acids in myoglobin. When correctly folded, the four subunits of hemoglobin stick together to form a shape like a pyramid. The subunits all have regions that allow them to adhere to each other strongly and precisely, in just the right orientation so that the right amino acids are in the right positions to do the right jobs.

The task hemoglobin has to do is trickier than myoglobin's. Myoglobin simply stores oxygen in muscles, but hemoglobin transports it from one place to another. To transport oxygen, hemoglobin not only has to bind the gas in the lungs where it is plentiful, it also has to release it to the peripheral tissues where it is needed. So it won't do for hemoglobin just to bind the oxygen tightly, since it then wouldn't be able to easily let it go where it was needed. And it won't do just to bind it loosely, because then it wouldn't efficiently pick up oxygen in the lungs. Like a Frisbee-playing dog that catches, brings back, and drops the saucer at your feet, hemoglobin has to both bind and release. Hemoglobin can bind oxygen tightly in your lungs and dump it off efficiently in your fingers and toes because of a Rube-Goldberg-like arrangement of the parts of the hemoglobin subunits. Here's a rough sketch of how it works. Don't worry about remembering the details—just notice the many precise mechanical steps.

When no oxygen is bound to hemoglobin, the iron atom of each subunit is a little too fat to fit completely comfortably into the hole in the middle of the heme where it resides. However, when an oxygen molecule comes along and binds to it, for chemical reasons the iron shrinks slightly. The modest slimming allows the iron to sink perfectly into the middle of the heme. Remember that "H" that was attached to the iron in myoglobin? (I knew you would!) Well, there also is an "H" attached in hemoglobin. As the iron sinks, it physically pulls along the attached "H." The "H" itself is part of one of the helical segments of the subunit, so when the "H" moves, it pulls the whole helix along with it. Now, at the interface of the subunits of hemoglobin, where alpha and beta chains contact each other, there are several positively charged amino acids across from negatively charged ones; of course they attract each other. But when the helix is pulled away by the "H" that's attached to the sinking iron, the oppositely charged groups are pulled away from each other (Figure A.2). What's more, the shape of the subunits is such that when one moves, they all have to move together. So hemoglobin changes shape into a somewhat distorted pyramid when oxygen binds, and electrostatic interactions between all of the subunits of hemoglobin are broken.

That takes energy. The energy to break all those electrical attractions comes from the avid binding of the oxygen to the iron. But here's the catch. Just as only one quarter dropped into the slot of a soda machine can't release the can, the binding of just one oxygen doesn't provide enough energy to break all those interactions. Instead, several subunits must each bind oxygen almost simultaneously to provide enough power. That only happens efficiently in a high-oxygen environment like the lungs. Conversely, when a hemoglobin that has four oxygen molecules attached to it is transported by the circulating blood from your lungs to the low-oxygen environment of, say, your big toe, when one of the oxygens falls off, the others aren't strong enough to keep the hemoglobin from snapping back. The electrostatic attractions between subunits reform, which yanks back the helix, which tugs up the "H," which pushes off the

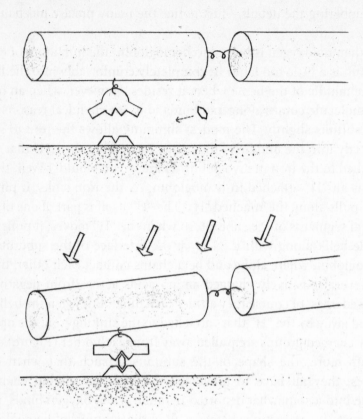

FIGURE A.2
Schematic drawing of how the binding of a small molecule can cause a protein to switch shape. (Top) Structural elements of a protein, represented by the cylinders and stippled surface, are apart. (Bottom) A small molecule, represented by the hexagon, binds to both surfaces, tilting the cylinders and bringing them closer to the stippled surface. The ability to switch its shape allows hemoglobin to deliver oxygen efficiently. (Drawing by Celeste Behe.)

oxygens. As a result, the remaining several oxygens are unceremoniously dumped off, exactly where they are needed.

My point in discussing the intricacies of the relatively simple molecular machine that is hemoglobin is not to tax the reader with details. Rather, the point is to drive home the fact that the machinery of the nanobot works by intricate physical mechanisms. Robots in our everyday, large-scale world (such as, say, robots in automobile factories that help assemble cars) function only if very many exactly shaped and precisely positioned parts—nuts, bolts, levers, wires, screws—are all in place and working. If they are ever built, artificial nanobots will also have to work by excruciatingly detailed physical mechanisms. Biological nanobots must do the same. There is no respite from mechanical complexity except in idle dreams or Just-So stories.

STICKING TOGETHER

Many molecular machines in the cell are much more complex than hemoglobin, but all work in the same mechanistic way. There are proteins that act as automatic gatekeepers, regulating the flow of small molecules or ions into and out of the cell. There are proteins that act as timing devices; others that are molecular trucks to ferry supplies to different parts of the cell; still others that act as cables and winches, pulling on cellular parts that need to be together. One of my favorites is a protein called gyrase, which can literally tie DNA into knots. In terms of our big, everyday world, gyrase is somewhat like a machine that could tie shoelaces. In developing an intuition for how such molecular machines act, a good start is to ask yourself how a shoelace-tying machine might work in our big world, or how a clock might work, or a delivery system, or a regulated gate. As you might suspect, they all would work by mechanical principles, and none of them would be simple.

Yet intuition can be insufficient. There is also a subtle but critical difference between molecular machines and everyday machines that needs to be kept in mind, a dissimilarity that underscores the much greater difficulty of making a molecular machine. One crucial way in which machinery in the nanobot differs from machinery in our everyday experience is that cellular machines have to *assemble themselves*. There is no conscious agent walking around in the cell putting pieces of machinery together, as there might be in a factory making, say, flashlights. Needless to say, the requirement for self-assembly enormously complicates the task of building a functional nanomachine.

How do cellular nanomachines build themselves? Here's a very simplified description. A protein binds to its correct partner(s) in the cell by having an area(s) on its surface that is closely complementary in shape and chemical properties with the other member(s) of the team. Let's think how that might work. Consider a protein with a positively charged amino acid on its surface. Of course the positive charge might attract a negative charge on the surface of another protein. However, there are thousands of different kinds of proteins in the cell and almost all have many negative charges. The interaction of just one positive and negative charge isn't enough to allow a protein to distinguish its partner from the many other proteins in the cell. So suppose that, next to the positive charge, the protein had an oily amino acid. Then it could match other proteins that had an oily patch next to a negative charge. Yet there will still be a lot of proteins in the cell with those two simple features, so even more specificity is needed. Further suppose next to the positive charge and oily patch there was a large amino acid sticking out from the surface. Then it could match a protein that had a negative charge, oily patch, and indentation in its surface. That combination of features decreases the number of potential partner proteins that it would match even further (see Figure 7.1).

In the last paragraph we worried about getting enough distinguishing features on the surface of a protein to allow it to discriminate between its correct binding partner and the thousands of other proteins in the cell that it should not stick to. But we also have to worry about the strength of the attraction. The reason is that a protein has to search blindly through the cell for its partner. It does so by randomly bumping into many surfaces, like pieces of flotsam and jetsam colliding with each other in a flowing stream, until the protein accidentally hits the complementary surface of its partner and sticks. However, suppose that the attraction between one positive and one negative charge were so overwhelmingly strong that whenever two opposite charges were close to each other they'd glom together, never to separate. If that were the case, the contents of the cell would congeal in an instant, killing it. The lesson is: Individual interactions can't be too strong. On the other hand, the total interaction strength of two proteins can't be too weak either, or the protein pieces might not form a stable entity, and might fall apart after a short time. The solution is to have a number of weak interactions between two proteins. Like Velcro fasteners, each individual interaction is rather delicate but the sum is strong. In the cell, multiple weak interactions make for strong binding. In general, the more interactions there are, the more specificity and strength there is to the binding between two proteins.

MAGNETS IN A SWIMMING POOL

As an illustration, imagine that a flashlight had to automatically self-assemble. To make the example closer to what happens in the cell, let's further imagine that the parts of the flashlight were floating in a big, well-stirred swimming pool, so that they could randomly bump against each other. Also imagine that thousands of other parts were floating in the pool, parts that belonged to other kinds of machinery. On the surface of all of the parts were tiny, rather weak, bar magnets, some with their north pole facing outward and south pole inward (buried inside the part, where it couldn't interact with other magnets), others with the south pole out and north in. As the water is stirred parts bump against other parts, some stick briefly when one or two magnets are in the right place to touch, but quickly break apart. When two pieces that are part of the future flashlight happen to collide in the correct orientation, they stick. The reason they stick, of course, is that there are multiple magnets (say, five to ten) on their surfaces in just the right positions, with just the right pattern, with exposed north poles arranged to be opposite exposed south poles. As with Velcro, the multiple weak interactions add up to a stable, strong binding. Then a third piece of the flashlight can stick to the growing conglomerate, and a fourth, until the flashlight is assembled. (Notice that the third and fourth pieces can't have the same pattern of magnets as the first and second pieces, or you wouldn't get the correct parts in the right order—for example, the battery might be stuck to the outside of the case!)

Let me make a few simple, interrelated points from this illustration. The first point is that of course parts of the flashlight all have to have patterns of magnets that match their binding partners. Put another way, even if all the correct pieces of the flashlight were floating in the pool, if none had magnet patterns to match each other, no flashlight would be made—the parts would occasionally bump, but wouldn't stick and thus wouldn't self-assemble into a flashlight. A further point is that the magnet features needed to form a binding pattern for a molecular machine to self-assemble are beyond the requirements for the function of the machine itself. In other words, the pattern of magnets that helps assemble the flashlight doesn't at all address the other aspects of the parts that allow them to act as a flashlight when assembled. Another point is that the binding patterns on a piece can't match incorrect parts. If the magnets on a piece of flashlight matched those on a piece of toaster, a mishmash would likely result, and would interfere with the construction of both flashlight and toaster.

A final, more subtle point is especially important for evaluating what Darwinian evolution can and can't do. Suppose we had a piece of one type of machine that we would also like to use in a different machine. Maybe we had a general-purpose part like a nut or bolt or gear. In our everyday world, of course, we could happily use the same type of nut or bolt in a thousand different machines. For example, a child's Lego building set can be used to make many different constructs. But when we're talking about self-assembling machinery, there's a major-league hitch. If a part has to attach to a partner different from its usual one, then the self-assembly instructions have to change. That is, the pattern of magnets on the surface of the part would have to be changed to match the new target. That might require multiple coherent changes before the part could assemble with the new target. What's more, if the assembly instructions changed, the part would lose its ability to assemble into the old system. To keep its old role while also gaining a new one, a near-duplicate of the old part would have to be made that had luckily acquired altered assembly instructions. For a process supposedly driven by random mutation, that would be a very tall evolutionary order.

Malaria Drug Resistance

In order to understand malaria's strengths, let's briefly look at a few examples of medicines that worked in ways different from chloroquine and are now being brushed aside by new mutations.

One set of treatments that was developed to take the place of chloroquine is abbreviated S/P, which stands for two different drugs, sulfadoxine with pyrimethamine. Both of these drugs target a vital metabolic pathway in malaria that builds components of DNA. It turns out that the four kinds of building blocks in DNA are of two types, called purines and pyrimidines. The parasite can obtain one type, purines, from the host it's invading, but has to make its own pyrimidines. So if its ability to make pyrimidines can be undercut, the bug is stymied. In order to make pyrimidines, the parasite, like other organisms, needs first to make several forms of a vitamin called folic acid. The two drugs in S/P, which both resemble natural chemicals in the metabolic pathway, block separate steps in the multistep pathway that makes pyrimidines. They do so by binding to the enzymes that normally catalyze the chemical conversions. However, mutations in the enzymes can make the drugs ineffective, probably by stopping them from binding. In the case of pyrimethamine (the "P" in the "S/P"), the drug interferes with an enzyme abbreviated DHFR. However, when a mutation appears in the enzyme and changes the amino acid at position number 108 from serine to asparagine, the drug loses its effectiveness. Similarly, when a mutation in an enzyme abbreviated DHPS changes the alanine normally found at position number 437 to a glycine, sulfadoxine (the "S" in the "S/P") fails.[1]

A hopeful note amid the gloom is that, about five years after the use of chloroquine was discontinued and S/P substituted in Malawi in 1993, the malaria there became susceptible to chloroquine once again. Some scientists have speculated that, if we're lucky, maybe drugs can be rotated; ineffective ones can be shelved for a while in the hope that they'll regain effectiveness sometime down the road.[2]

A relatively new drug, atovaquone, which interfered with a different step in *P. falciparum* metabolism, can be countered by a single amino acid mutation in a protein called cytochrome *b*.[3] The very latest drug, artemisinin, is derived from the Chinese sweet wormwood plant. Resistance to artemisinin has not yet been seen in clinics, but has been reported in laboratory investigations, and will almost certainly develop in the field eventually.[4] Nicholas White of Mahidol University in Thailand worries, "If we lose artemisinins to resistance, we may be faced with untreatable malaria."[5] Quinine, the natural drug that first turned the tide of battle toward humanity's side, is still pretty effective against *P. falciparum*. But the bug is slowly gaining ground, apparently by many little changes in a number of separate genes (like sickle hemoglobin and HPFH on the human side) rather than in one gene, as for chloroquine resistance.[6]

Assembling the Bacterial Flagellum

THE OUTBOARD MOTOR

The cilium is an elegant molecular machine that powers the swimming of cells as diverse as sperm and pond algae. As we've seen, not only is the cilium itself enormously complex, but IFT—the system that builds the cilium—is also highly sophisticated, intricate, and dynamic. Without the assembly system, no working machinery gets built. The need to spontaneously assemble intricate machinery enormously complicates any putative Darwinian explanation for the foundation of life, which has to select from tiny, random steps the size of the sickle cell mutation. Yet IFT is not some fantastic aberration. In a cellular nanobot, where machines run the show without the help of conscious agents, *everything* has to be assembled automatically. To drive home the complexity of self-assembly, let's look at just one more example—the bacterial flagellum.

The flagellum is a cellular propulsion system that is completely different from the cilium. Rather than acting as an oar that goes back and forth like the cilium, the flagellum is a rotary motor—literally an outboard motor that bacteria use to swim. And just like the familiar outboard motor that powers a boat on a lake, the flagellum needs many different parts to work. Although it consists of dozens of different protein parts, when I wrote about the flagellum in *Darwin's Black Box* I focused on just the several mechanical parts—propeller, motor, and stator—that all rotary motors need to work, to show the system was irreducibly complex. As one biochemistry textbook put it, the bacterial rotary motor "must have the same mechanical elements as other rotary devices: a rotor (the rotating element) and a stator (the stationary element.)"[1]

However, not all rotary devices are equal. For example, the rotary device that spins the wheels on my son's toy car is a far cry from the kind that operates a real motorboat. In turn, the motorboat's engine is quite different in many details from one that powers an ocean liner. These different rotary systems all have a large number of parts—not just two or three—all of which are necessarily precision-machined to the right shapes for the job. If one were to try to realistically sketch out the kind of automated assembly machinery that would put together any one of these, it would be quite different from the assembly machinery for any other of them. The assembly machinery would have to be different because the details of the assembly itself are different—the distance that one newly made part is from another in the staging area, which nut goes onto which bolt, what size clamp is needed to grasp a part, and so on. So when we are thinking about the assembly of the flagellum, we have to think about all the specific details of the particular machinery we're making. Let's briefly consider the structure of the bacterial flagellum.

Figure C.1 shows a sketch of a flagellum taken from a recent article in a science journal describing how a flagellum is built.[2] A flagellum contains several dozen different kinds of protein parts, many of which are labeled with their scientific names in Figure C.1. The labels give a taste of the complexity of the parts, but in the following description I won't use those labels—I'll use more reader-friendly terms. Again, don't think you have to memorize the details—just taste the complexity.

A flagellum can be conceptually broken down into three subsystems: the base (which contains the motor), the "hook" (which acts as a universal joint), and the filament (which is the propeller). Within each subsystem, however, are multiple precision-made parts. The base contains the motor that drives the rotation of the flagellum. It also contains protein parts that act as the stator (to clamp the structure firmly in place), as well as bushings and a protein pump that, as we'll see below, is critical to the assembly of the flagel-

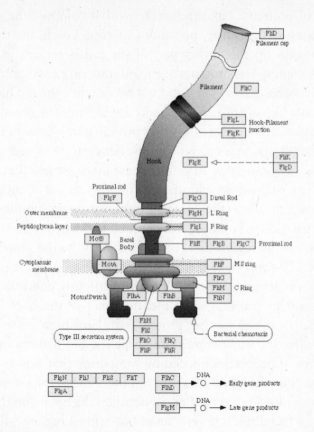

FIGURE C.1

The bacterial flagellum. Protein components of the system are labeled in detail. (Reprinted courtesy of the Kyoto Encyclopedia of Genes and Genomes, Kanehisa, M., Goto, S., Hattori, M., Aoki-Kinoshita, K. F., Itoh, M., Kawashima, S., Katayama, T., Araki, M., and Hirakawa, M. 2006. From genomics to chemical genomics: new developments in KEGG. *Nucleic Acids Res.* 34:D354–57.)

lum. The structure of the base is made of several rings, one of which (the MS ring) is in the cell membrane, the next of which (the P ring) is in the cell wall, and the next of which (the L ring) is in the outer membrane. Each of the three rings is made up of about twenty-six copies of its particular protein component.

As shown in Figure C.1, through the rings is placed a rod, which acts as the drive shaft for the flagellum, transmitting the rotation of the motor to the filament-propeller. The rod contains several different kinds of proteins. The three proteins that compose the part of the rod closest to the cell are present in six copies each, and the protein that makes up the farther part of the rod is present in about twenty-six copies. The proteins of the interior ring have cylindrical symmetry, like balls arranged on a hula hoop, while the rod has helical symmetry, like the thread of a wood screw. Since the two symmetries are mismatched, there is another protein part—present in nine copies—that seems to act as an adaptor between them, reconciling the discordant symmetries. Also in the base is the protein that acts as the motor, as well as three kinds of proteins that act as molecular switches, which allow the motor to change from spinning in a clockwise rotation to spinning in a counterclockwise one.

The hook is the region that connects the base to the propeller. It consists of 120 copies of another type of protein. When it is being assembled, the length of the hook has to be tightly controlled so it isn't too short or too long. The measurement of the hook length seems to be the job of another protein part. How it measures is not yet clear. After the hook comes the propeller. But it turns out that the mechanical properties needed by something that acts as a universal joint (like the hook does) are not the same as the mechanical properties needed for a propeller. So between the hook and the propeller in the flagellum is a very small but critical region called the junction zone, where several other protein parts (present in copies of a baker's dozen apiece) act as adaptors to fit the two disparate pieces together. In other words, "It seems very likely that the junction zone acts as a buffering structure connecting two filamentous structures with distinct mechanical characteristics."[3]

The propeller itself is made of tens of thousands of copies of flagellin, a sophisticated protein that can switch between several different shapes. The different shapes then give the elongated propeller a different curl, with varying swimming properties. Although the word "flagellum" comes from the Latin for "whip," the propeller turns out not to be a solid structure like a bullwhip. Instead, it's hollow like a drinking straw. This feature is critical for the assembly of the flagellum, as we'll now see.

BUILDING THE OUTBOARD MOTOR

In just the past ten years or so, through the hard work of many scientists in many labs in many countries, details of how a flagellum is built in a bacterial cell have been pieced together. Although many aspects remain hazy, enough is now clear to give a fascinating overview of the elegance and complexity of the assembly process. An animation of the construction of a flagellum has been produced by the "Protonic NanoMachine Project" of the Japan Science and Technology Corporation in a remarkable video, which can be viewed on the Web.[4]

Like the cilium and the tower at Iacocca Hall, the flagellum is built from the bottom up. The first component to be laid down is the basement—the protein ring in the inner cell membrane (the MS ring). Then, using that structure as a foundation, a sort of housing unit is built on the inside of the cell (called the C ring). Inside the housing is then assembled a machine, called a Type III export apparatus. The export machinery is like a gun that grabs the correct proteins (which are suitably labeled so the automated machinery can distinguish them from proteins that are not part of the flagellum) and pushes them out to the end of the growing structure. The first proteins to be pushed through are those that make up the rod, along with a special protein that can chew through the cell wall. That is needed so the flagellum can grow beyond the stiff boundary of the cell.

The next stage is the assembly of the other rings, L and P. The proteins that make up these structures don't come through the regular way, however; they are pushed out of the cell by a different set of machinery that is used for the secretion of a variety of other proteins. The protein that makes up the P ring can't get to the incipient flagellum by itself—it needs another protein called a chaperone to shepherd it over to the construction site; otherwise, the protein loses its way and never arrives. After escorting the P ring protein to its proper destination, the chaperone floats away.

Once the rod is finished, another protein is pushed through the middle of the growing structure to start the hook. The protein isn't one that will be part of the final structure, however. Rather, it's called the "hook cap" protein; it helps keep the actual building components in place as the flagellum grows. After the hook is assembled, the hook cap falls off and floats away. The proteins that make up the junction zone are then grabbed by the export machinery and sent through the export channel to the end of the nascent flagellum.

Finally, we're just about ready to start the business end of the flagellum, the propeller that actually pushes the bacterium forward. But before we do, there's another critical step. Just as the construction of the hook region needed a "cap" at the end, so does the propeller. But it's not the hook cap; it's a different cap. So before the protein pieces that make up the propeller are sent through the export machinery, a "filament cap" precedes them. The cap fits on the end of the hollow flagellum, and as each of the tens of thousands of copies of the propeller protein are pushed down the center to the end, the cap prevents them from spewing out into the surrounding liquid and being lost. In order to traverse the rather thin, hollow central channel of the flagellum, the flagellar proteins have to be kept in an extended shape. When they arrive at the far end, the cap also helps all the copies of the propeller protein to fold into the correct, compact shape—the shape needed to form the propeller.

THE BALLERINA

While describing the structure of the flagellum in this section I've written rather blandly of "protein parts" for this and that, as if the individual proteins were like so many simple nuts and bolts. That is not the case at all. Like hemoglobin, all of the dozens of proteins involved in building the flagellum are themselves quite intricate and wonderfully suited to their jobs. To drive home the point, for illustration let's look at just one example—the filament cap.

The filament cap is made up of five copies of a single protein whose official name is "FliD" but I'll call it "Twinkletoes." For comparison, remember that hemoglobin has four parts—two alpha and two beta chains. When stuck together, the five protein parts give Twinkletoes a shape that might best be described as a starfish on stilts. The leglike stilts point vertically down from the horizontal pentagonal starfish. Now, the hollow filament of the flagellum is made of multiple copies of flagellin protein arranged in eleven strands, so the fivefold symmetrical cap is slightly mismatched to the ends of the filament. One leg of the cap can fit in a crease between every other strand, but two times five is ten, not eleven, so one crease does not have a cap leg stuck in it.

But the mismatch is not some mistake; it's part of the elegant design of the assembly system. As a copy of flagellin protein is pushed down the hollow tube to be added to the growing end of the filament, it is prevented from floating out into space by the filament cap. The cap allows the flagellin time to fold to its functional shape, and then directs it to fill the empty space on the growing filament. So the "mismatch" actually directs the protein to the correct, available position. As the flagellin fills the proper vacant position, the pentagonal cap rotates, so that the next available slot is now in position to be filled. To do this, Twinkletoes lifts one of its legs and moves it over a notch. The next copy of flagellin then comes down the follow tube of the filament and is directed to the right spot, Twinkletoes rotates again to the next space, and the next leg swings over. Tens of thousands of times the dancing machinery[5] automatically directs the right building blocks to the right positions, lifts its supple legs, and spins to the next position.

ANOTHER MATTER

How do Darwinists explain the flagellum? In the same way as they explain the cilium—usually by a tactful silence, occasionally by Just-So stories. There is currently a lively discussion going on in the professional science literature about the flagellum and another structure called a "Type III secretory system" (TTSS), which contains a number of protein parts that resemble those of the flagellum. The TTSS is used by bacteria as a protein pump; since parts of the flagellum also act as a pump in order to build the flagellum, some workers reasonably think that the two are related by common descent. Whether the TTSS or the flagellum came first is the point of controversy.[6] But *none* of the papers seriously addresses how either structure could be assembled by random mutation and natural selection, or even how one structure could be derived from the other by Darwinian processes.[7] Consider a review of flagellar assembly written by the eminent Yale biologist Robert Macnab shortly before his premature death in 2003. The article of course shows great erudition, and it nicely summarizes the startlingly complex pathway of flagellum assembly.

How did such a pathway evolve by random mutation? In the approximately seven-thousand-word review, the phrase "natural selection" does not appear. The word "evolution" or any of its derivatives occurs just once, in the very last sentence of the article. Speaking of the flagellum and the TTSS, Macnab writes: "Clearly, nature has found two good uses for this sophisticated type of apparatus. How [the TTSS and the flagellum] evolved is another matter, although it has been proposed that the flagellum is the more ancient device, since it exists in bacterial genera that diverged long before eukaryotic hosts existed as virulence targets."

Darwinism has little more of substance to say.

The Cardsharp

STACKING THE DECK

One intriguing possibility for Darwinian construction of cellular machines that has been much discussed in the scientific literature recently is the shuffling around of binding sites, to bring different proteins close to one another.[1] To illustrate, suppose there were two large pegboards on the wall of a carpentry shop, with chalk outlines drawn of which tools were supposed to be hung on which pegs, with a different set of tools on each of the two different pegboards. If we cut the two pegboards down the middle and switched two halves of the two boards, we'd have different tools next to each other than we had before, without having to draw a new outline of a tool in a new position.

Something like that is thought to explain some features of cells of higher organisms (eukaryotic cells). Some proteins resemble several proteins that have been stitched together. Such proteins have discrete regions called "domains"[2] that can each fold up into compact shapes, the way myoglobin does. The domains are often connected by short, thin lengths of the amino acid sequence of these multidomain proteins. The thin lengths look like they do little more than just tie the domains together. In some proteins, several or all of the domains have binding sites for other proteins, with a different kind of protein binding to each domain. The apparent purpose of these particular multidomain proteins is just to bring the other proteins close together (Figure D.1).

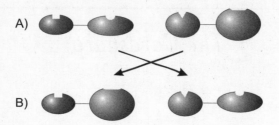

FIGURE D.1
Cartoon illustrating "domain swapping" of proteins. A) Two proteins, each consisting of two linked domains. Each domain has a binding site for a separate, different protein, indicated by the differently shaped depressions. B) Mutational processes rearrange the genes for the proteins, generating novel combinations of binding sites.

What do such so-called "scaffold" proteins do? Fascinatingly, many seem to act as little computer circuits, signaling a cell to make appropriate "decisions" in a changing world. The cell is an extremely complex system that has to respond suitably to a variety of circumstances. It has to grow at the right time, defend itself when necessary, search for food, even self-destruct sometimes for the greater good. To be able to do all of this, the cellular nanobot has to collect information about the environment, weigh it, and then use the information to take effective action. As one group of scientists notes, "Cells require a remarkable array of sophisticated signal processing behaviors that rivals or surpasses that of modern computers."[3] So one scaffold protein might have binding sites for proteins that indicate some condition (like, say, it's time to grow now) as well as binding sites for proteins that will then take the appropriate action (like, say, sending a definitive signal to the nucleus to start replicating). Another scaffold protein might have binding sites for proteins that tell a cell to kill itself (perhaps sent by immune cells that "perceive" the doomed cell has been invaded by a virus) as well as proteins that begin the autodestruct sequence.

Scaffold proteins have been likened to parts of computer programs[4] called "AND" gates or "OR" gates.[5] It's common for a human
programmer to write some computer code that in English says "IF
(one condition is true) AND (another condition is true) THEN
(execute action number one)." A scaffold protein that conveys a
certain signal only IF one certain protein AND another particular
protein are bound to it is acting like that computer statement. A
programmer might also write "IF (one condition is true) OR
(another condition is true) THEN (execute action number two)." A
scaffold protein that conveys a signal IF either one certain protein
OR another particular protein is bound is acting like that computer
statement. As one might imagine,[6] more complex computer or protein circuits could easily be generated.

Suppose, however, that the two computer statements got mixed
up. Suppose that through some glitch we got the statement "IF (one
condition is true) AND (another condition is true) THEN (execute
action number *two*)." If the statements somehow got mixed up, the
input conditions for the first AND statement would be linked to the
output condition for the second OR statement. If something like
that happened for two scaffold proteins, a new biological circuit
might be made without having to produce any new protein-binding
sites.

Exactly that scenario has been modeled by the group of Wendell
Lim, a biologist at the University of California at San Francisco.[7]
Using clever laboratory techniques, in one experiment Lim and
coworkers spliced a yeast scaffold protein that normally binds a
protein that allows the yeast to mate with a second protein that
receives a signal that tells the yeast to brace itself against extra-salty
water. As hoped, the result was a new signaling circuit—yeast that
had the hybrid protein could only survive in concentrated salt solu-
tions in the laboratory if they were exposed to the mating signal
protein. In another set of experiments[8] Lim's group constructed
artificial proteins using multiple different domains. One of the
domains regulated the formation of actin fibers; the other domains
bound various other proteins. Under some conditions in the test
tube the artificial scaffold proteins either didn't work at all or were
turned on all the time. But in other conditions some proteins could

act as either an AND circuit or an OR circuit, just as the scientists planned.

Lim thinks such results will help us both to engineer cells and to understand evolution:

> These findings demonstrate that scaffolds are highly flexible organizing factors that can facilitate pathway evolution and engineering. . . . [P]rimitive tethering scaffolds generated by recombination or fusion events could *in principle* [emphasis added] be sufficient to generate new pathways. . . . [T]hese organizing structures thus appear to be optimized for evolvability, a property that may provide increased fitness in the face of constantly changing environmental challenges and signaling needs. . . . [S]caffold engineering may allow for systematic manipulation of cytoplasmic signaling pathways.

Although the results do show great promise for the productive engineering of cells by intelligent agents, I do not believe they indicate that an incoherent process could build new, complex, helpful genetic circuits by randomly rearranging old parts. The simple point that even superb scientists like Lim—who assume a Darwinian framework—do not seem to grasp is that the purposeful arrangement of parts (including by scientists in laboratories) is the hallmark of intelligence. It does not mimic random mutation. It is the exact opposite of random mutation.

Lim of course doesn't claim his work is an actual example of evolution in action, but he does view it as a sort of proof of principle that such a phenomenon is theoretically possible. So it's worth recalling the key insight of evolutionary biologists Jerry Coyne and Allen Orr that "the goal of theory, however, is to determine not just whether a phenomenon is theoretically possible, but whether it is *biologically reasonable*—that is, whether it occurs with significant frequency under conditions that are likely to occur in nature." What do the lab results tell us about whether random-yet-productive shuffling of domains "occurs with significant frequency under conditions that are likely to occur in nature"? About whether that is *biologically reasonable*? Nothing at all. When a scientist intentionally arranges fragments of genes in order to maximize the chances of their interacting productively, he has left Darwin far, far behind. You don't learn much about the fair odds of winning at poker by watching a cardsharp deal himself a royal flush, and you don't learn much about random mutation by arranging genes in the lab on purpose.

DOMAIN SHUFFLING IN THE REAL WORLD

On which side of the edge of evolution would domain shuffling be expected to fall in nature, rather than in the lab? Is it biologically reasonable to think that random mutation and natural selection could build new, coherent genetic circuits from old protein-binding sites? One big difficulty in coming to a firm conclusion on that question is that, unlike the situation with respect to shape space and protein-protein binding sites, there have been no good experiments that show what fraction of mutations would work—nothing like the experiments of Greg Winter's lab and others that showed that between ten and a hundred million binding sites have to be searched in a shape space library to find one that will bind with a modest affinity to a second protein. So any conclusion we reach will be less quantitative and more tentative than for the development of brand new protein-protein binding sites.

Nonetheless, there is information available that can help us make an informed judgment. First, in all of the experiments, Lim's lab didn't just splice two genes together in a single step; they took several additional steps as well. For example, in the case of the hybrid mating factor/concentrated salt scaffold protein they added further mutations to knock out the original pathway, to ensure there was no cross-reaction where, say, one signal would activate both the mating response and the high salt response. Remember, the more steps that have to occur between beneficial states, the much less plausible are Darwinian explanations.

Second, in joining together various protein-binding domains to control actin assembly, Lim's group found quite complex results:

Switches could be divided into diverse behavioral classes. At the extremes, five switches showed little or no basal repression, and nine were extremely well-repressed, but could not be activated under any of the tested conditions. Most constructs, however, showed some type of gating behavior. . . . Heterologous switch behavior was also dependent on affinity of the autoinhibitory interactions. . . . Linker length also affected switch behavior. . . . [I]ncreasing interdomain linker length did not uniformly reduce coupling, which suggests that these effects are context-dependent. . . . The combinatorial switch library also yielded switches with the unexpected behavior of antagonistic or negative input control. . . . This unanticipated class of switches highlights a striking feature of the library: Subtle changes in switch parameters can lead to dramatic changes in gating behavior.[9]

In other words, the system behavior is chaotic and incoherent, depending on many conflicting factors. Which of the various possibilities would be harmful to an organism? Which of the very few that might be helpful for the moment would be evolutionary dead ends, single steps to local peaks in a rugged evolutionary landscape? In the mating/salt tolerance experiment, the poor mutated yeast was sterile, unable to mate, and could only resist high salt concentrations if supplied with mating factor. To say the least, such a response would be unlikely to help in nature.

The third and most important factor in judging how helpful domain shuffling is likely to be is that P. falciparum seems to have made no use of it. In a hundred billion billion chances, when the malarial parasite was in a life-or-death struggle with chloroquine, domain shuffling was nowhere to be seen. Writes Lim: "By allowing the establishment of novel regulatory connections between molecules with no previous physiological relation, such recombination events would be a powerful force driving evolution of novel cellular circuitry." Yet the fancied "powerful force" wasn't as helpful as a few, simple, run-of-the-Darwinian-mill point mutations in PfCRT.

Domain shuffling would be an instance of the "natural genetic engineering" championed by James Shapiro, where evolution by big random changes is hoped to do what evolution by small random mutations can't. But random is random. No matter if a monkey is rearranging single letters or whole chapters, incoherence plagues every step. Although we have a less secure quantitative base for deciding, and new data might bear on the question one way or another, it's likely that domain rearrangement is similar to everything else that random mutation does. One step might luckily be helpful on occasion, maybe rarely a second step might build on it. But Darwinian processes in particular and unintelligent ones in general don't build coherent systems. So it is biologically most reasonable to conclude that, like multiple brand-new protein-protein binding sites, the arrangement of multiple genetic elements into sophisticated logic circuits similar to those of computers is also well beyond the edge of Darwinian evolution.

COMPUTER ASSUMPTIONS

What about computers themselves, though? If some aspect of biology can be mimicked accurately on a computer, wouldn't that allow us to probe the edge of evolution in greater detail? In principle, it would. The problem is that living things are so complex that all descriptions of them, whether in computers or books, require the kind of drastic simplification that can lead to serious error if we're not careful. A prominent example is Avida, an "artificial life" computer program that, according to its inventors, explains "how complex functions can originate by random mutation and natural selection." [10]

In Avida, an "organism" is a sequence of computer instructions coupled with a processor that executes these instructions in sequence. Just as we burn calories with every activity we engage in, these artificial organisms burn computational "energy" with each instruction executed. They, like us, have to feed themselves if they want to survive. In Avida, these artificial organisms are awarded extra computer "food" if they manage to acquire a set of instructions that performs a simple computational task. (Let's not worry about the computer details of how instructions are acquired or lost.) And, as you may have guessed, random mutation and natural selection seem to be perfectly capable of delivering the needed instruction sets.

What are we to make of this apparent contradiction? If, as we have seen, random mutation is incoherent and severely constrained in our best evolutionary studies of real biological organisms, how can a process that is supposedly analogous to Darwinism work for a computer program? The simple answer is that the conclusions drawn from an analogy are only as good as the analogy. Although Avida is lifelike in a few respects, it only takes one critical departure for the overall analogy to fail.

Let's look at just one example to illustrate the point. In Avida, acquiring new abilities is only one way for an organism to get computer food. Another way is by simply acquiring surplus instructions, *whether or not they do anything*. In fact, instructions that aren't ever executed—making them utterly useless for performing tasks— are beneficial in Avida because they provide additional food without requiring any additional consumption. It's survival of the fattest!

It's also very unrealistic. Biological organisms show the opposite behavior—genes that are useless in the real world are not rewarded; the genes are rapidly lost or degraded by mutation. Why, then, was Avida programmed to do the opposite—to reward organisms for carrying useless instructions? As explained on the Avida website, the counterbiological feature was needed, "Otherwise there is a strong selective pressure for shorter genomes."[11] In other words, otherwise the program wouldn't give the desired results. The computer programmers remark, "This isn't the most elegant fix, but it works."

Computers can be useful tools in science when the assumptions built into programs are realistic. But if assumptions are wrong, computer simulations can be misleading. That's why the most informative evolutionary studies by far are ones of real organisms such as malaria. The million-murdering death makes no assumptions.

Notes

---

2 Arms Race or Trench Warfare?

1. The phrase is from a poem found in a letter written by the British scientist Ronald Ross to his wife on August 20, 1897, which he called "Mosquito Day." Ross (1857–1932), who discovered that malaria is transmitted by mosquitoes, won the 1907 Nobel Prize in Physiology or Medicine. The poem is reproduced in Sherman, I. W. 1998. *Malaria: parasite biology, pathogenesis, and protection*. Washington, D. C., ASM Press, 6.

2. Born into an affluent Grenadian family, the well-educated Noel came to the United States in 1904 to study dentistry at the Chicago College of Dental Surgery. Despite occasional hospitalizations for the effects of his unrecognized sickle cell disease, Noel graduated in 1907 and returned to his native Grenada, where he set up a successful dental practice. He died at the age of thirty-two from "asthenia from pneumonia," probably as a secondary result of sickle cell disease. Savitt, T. L., and Goldberg, M. F. 1989. Herrick's 1910 case report of sickle cell anemia. The rest of the story. *JAMA* 261:266–71.

3. Pauling, L., Itano, H. A., Singer, S. J., and Wells, I. C. 1949. Sickle cell anemia, a molecular disease. *Science* 110:543–48.

4. Ingram, V. M. 1958. Abnormal human haemoglobins. I. The comparison of normal human and sickle-cell haemoglobins by fingerprinting. *Biochim. Biophys. Acta* 28:539–45; Hunt, J. A., and Ingram, V. M. 1958. Abnormal human haemoglobins. II. The chymotryptic digestion of the trypsin-resistant core of haemoglobins A and S. *Biochim. Biophys. Acta* 28:546–49.

5. Although some authors think the sickle gene arose independently more than once, Cavalli-Sforza argues for a single origin (Cavalli-Sforza, L. L., Menozzi, P., and Piazza, A. 1994. *The history and geography of human genes*. Princeton, N.J.: Princeton University Press.)

6. Forget, B. G. 1998. Molecular basis of hereditary persistence of fetal hemoglobin. *Ann. N.Y. Acad. Sci.* 850:38–44.

7. Bookchin, R. M., Nagel, R. L., and Ranney, H. M. 1967. Structure and properties of hemoglobin C-Harlem, a human hemoglobin variant with amino acid substitutions in 2 residues of the beta-polypeptide chain. *J. Biol. Chem.* 242:248–55.

8. Except under extreme conditions not found in a normal red blood cell.

9. Let's compare one population in which half of the genes are normal and half sickle to another population in which half of the genes are normal and half C-Harlem. Assuming random inheritance, one-quarter of both populations would have two normal genes and be vulnerable to malaria, and one-half of both populations would have one normal and one mutant gene, which in both cases would confer resistance to malaria. Two copies of sickle are for all intents and purposes lethal. Two copies of C-Harlem would not be lethal in themselves, but might not provide much protection against malaria. So the remaining quarter of the first population would have two copies of the sickle gene and die, presumably before reproducing. The remaining quarter of the second population would have two copies of the C-Harlem gene and live for a while, but with presumably little or no resistance to malaria. Assuming their vulnerability to malaria leads to a 50 percent mortality rate before reproducing, then the advantage of the second population would be in the one-half of the quarter of the population with two copies of the mutant gene that managed to survive malaria. This amounts to a one-eighth selective advantage of C-Harlem over sickle—a very large edge in evolutionary terms.

A seemingly even better solution to the problem hasn't turned up yet in nature. If a normal beta gene and a sickle gene could be brought together on the same chromosome (similar to something like Hb anti-Lepore: Efremov, G. D. 1978. Hemoglobins Lepore and anti-Lepore. *Hemoglobin* 2:197–233) and be equally expressed, then a population in which that arrangement was universal would be superior to either of the two mixed populations discussed above. If everyone in the population expressed both normal and sickle hemoglobin, then one would effectively have a population where everyone had sickle trait, with no cases of either sickle disease (lethal in itself) or vulnerability to malaria.

10. Tishkoff, S. A., Varkonyi, R., Cahinhinan, N., Abbes, S., Argyropoulos, G., Destro-Bisol, G., Drousiotou, A., Dangerfield, B., Lefranc, G., Loiselet, J., Piro, A., Stoneking, M., Tagarelli, A., Tagarelli, G., Touma, E. H., Williams, S. M., and Clark, A. G. 2001. Haplotype diversity and linkage disequilibrium at human G6PD: recent origin of alleles that confer malarial resistance. *Science* 293:455–62.

11. Another interesting hemoglobin mutation that I won't discuss is called

hemoglobin E. Like HbC, it has a glutamic acid to lysine change in the beta chain. However, the position in the beta chain that's altered is different. In HbE, amino acid number 26 is changed. HbE appears mostly in Asian populations and is thought to have antimalarial properties because it is found frequently in malarious regions.

12. Modiano, D., Luoni, G., Sirima, B. S., Simpore, J., Verra, F., Konate, A., Rastrelli, E., Olivieri, A., Calissano, C., Paganotti, G. M., D'Urbano, L., Sanou, I., Sawadogo, A., Modiano, G., and Coluzzi, M. 2001. Haemoglobin C protects against clinical *Plasmodium falciparum* malaria. *Nature* 414:305–8.

13. The threat of death from malaria diminishes for those who are constantly exposed to it—not for adults who have only infrequently or never been exposed.

14. Darwin married his cousin Emma, with whom he had ten children.

15. Children with one C gene and one sickle gene have symptoms similar to, but milder than, sickle cell disease.

16. Some things that could greatly complicate the outcome include: the development of other hemoglobin mutations (like C-Harlem); occurrence of other, nonhemoglobin, mutations in humans; the rate of flow of normal hemoglobin genes (HbA) into the populations; the effects of inbreeding; changes in the malarial parasite *P. falciparum*; and changes in the mosquitoes that carry the parasite.

17. Fortin, A., Stevenson, M. M., and Gros, P. 2002. Susceptibility to malaria as a complex trait: big pressure from a tiny creature. *Hum. Mol. Genet.* 11:2469–78; Kwiatkowski, D. 2000. Genetic susceptibility to malaria getting complex. *Curr. Opin. Genet. Dev.* 10:320–24; Mazier, D., Nitcheu, J., and Idrissa-Boubou, M. 2000. Cerebral malaria and immunogenetics. *Parasite Immunol.* 22:613–23; Kwiatkowski, D. P. 2005. How malaria has affected the human genome and what human genetics can teach us about malaria. *Am. J. Hum. Genet.* 77:171–92.

18. Strachan, T., and Read, A. P. 1999. *Human Molecular Genetics,* 2nd ed. Wiley: New York, Table 17.2.

19. The alpha+ thalassemias. (Carter, R., and Mendis, K. N. 2002. Evolutionary and historical aspects of the burden of malaria. *Clin. Microbiol. Rev.* 15:564–94; Allen, S. J., O'Donnell, A., Alexander, N. D., Alpers, M. P., Peto, T. E., Clegg, J. B., and Weatherall, D. J. 1997. Alpha+-thalassemia protects children against disease caused by other infections as well as malaria. *Proc. Natl. Acad. Sci. USA* 94:14736–41; Flint, J., Hill, A. V., Bowden, D. K., Oppenheimer, S. J., Sill, P. R., Serjeantson, S. W., Bana-Koiri, J., Bhatia, K., Alpers, M. P., and Boyce, A. J. 1986. High frequencies of alpha-thalassaemia are the result of natural selection by malaria. *Nature* 321:744–50.

20. Greene, L. S., and Danubio, M. E. 1997. *Adaptation to malaria: the*

interaction of biology and culture. Gordon and Breach Publishers: Amsterdam.

21. The gene for G6PD occurs on the X chromosome. Women have two X chromosomes but men have only one (in addition, they have a Y chromosome). So a woman can have a broken G6PD gene on one of her X chromosomes but a working copy on the other. If a man inherits a broken G6PD gene, he has no second copy to back it up.

22. Ruwende, C., Khoo, S. C., Snow, R. W., Yates, S. N., Kwiatkowski, D., Gupta, S., Warn, P., Allsopp, C. E., Gilbert, S. C., and Peschu, N. 1995. Natural selection of hemi- and heterozygotes for G6PD deficiency in Africa by resistance to severe malaria. *Nature* 376:246–49.

23. Alberts, B. 2002. *Molecular biology of the cell,* 4th ed. New York: Garland Science, pp. 604–5.

24. Carter and Mendis. 2002.

25. Kennedy, J. R. 2002. Modulation of sickle cell crisis by naturally occurring band 3 specific antibodies—a malaria link. *Med. Sci. Monit.* 8:HY10-HY13.

26. Pogo, A. O., and Chaudhuri, A. 2000. The Duffy protein: a malarial and chemokine receptor. *Semin. Hematol.* 37:122–29. The nucleotide at position –33 in the gene is changed from a T to a C. The mutation abolishes the binding site for h-GATA-1 erythroid transcription factor.

27. Dawkins, R. 1986. *The blind watchmaker.* New York: Norton, p. 178.

28. Ibid., p. 181.

3   The Mathematical Limits of Darwinism

1. Hastings, I. M., Bray, P. G., and Ward, S. A. 2002. Parasitology. A requiem for chloroquine. *Science* 298:74–75.

2. Plowe, C. V. 2005. Antimalarial drug resistance in Africa: strategies for monitoring and deterrence. *Curr. Top. Microbiol. Immunol.* 295:55–79.

3. Wellems, T. E., Walker-Jonah, A., and Panton, L. J. 1991. Genetic mapping of the chloroquine-resistance locus on *Plasmodium falciparum* chromosome 7. *Proc. Natl. Acad. Sci. USA* 88:3382–86.

4. Su, X., Kirkman, L. A., Fujioka, H., and Wellems, T. E. 1997. Complex polymorphisms in an approximately 330 kDa protein are linked to chloroquine-resistant *P. falciparum* in Southeast Asia and Africa. *Cell* 91:593–603.

5. Gardner, M. J., et al. 2002. Genome sequence of the human malaria parasite *Plasmodium falciparum. Nature* 419:498–511.

6. Bray, P. G., Martin, R. E., Tilley, L., Ward, S. A., Kirk, K., and Fidock, D. A. 2005. Defining the role of PfCRT in *Plasmodium falciparum* chloroquine resistance. *Mol. Microbiol.* 56:323–33.

7. Fidock, D. A., Nomura, T., Talley, A. K., Cooper, R. A., Dzekunov, S. M., Ferdig, M. T., Ursos, L. M., Sidhu, A. B., Naude, B., Deitsch, K. W., Su, X. Z., Wootton, J. C., Roepe, P. D., and Wellems, T. E. 2000. Mutations in the *P. falciparum* digestive vacuole transmembrane protein PfCRT and evidence for their role in chloroquine resistance. *Mol. Cell* 6:861–71.

8. Two different origins in South America, one in Papua–New Guinea, and one in Asia that spread to Africa. (Wootton, J. C., Feng, X., Ferdig, M. T., Cooper, R. A., Mu, J., Baruch, D. I., Magill, A. J., and Su, X. Z. 2002. Genetic diversity and chloroquine selective sweeps in *Plasmodium falciparum. Nature* 418:320–23.)

9. Kublin, J. G., Cortese, J. F., Njunju, E. M., Mukadam, R. A., Wirima, J. J., Kazembe, P. N., Djimde, A. A., Kouriba, B., Taylor, T. E., and Plowe, C. V. 2003. Reemergence of chloroquine-sensitive *Plasmodium falciparum* malaria after cessation of chloroquine use in Malawi. *J. Infect. Dis.* 187:1870–75; Cooper, R. A., Hartwig, C. L., Ferdig, M. T. 2005. *Pfcrt* is more than the *Plasmodium falciparum* chloroquine resistance gene: a functional and evolutionary perspective. *Acta. Trop.* 94:170–80. Drug resistance mutation in pfmdr, the other protein involved in chloroquine resistance, also incurs a fitness cost (Hayward, R., Saliba, K. J., Kirk, K. 2005. pfmdr1 mutations associated with chloroquine resistance incur a fitness cost in *Plasmodium falciparum. Mol. Microbiol.* 55:1285–95).

10. Nourse, A. E. 1963. Tiger by the Tail. In *Fifty Short Science Fiction Tales,* eds. I. Asimov and G. Conklin. New York: Collier Books. pp. 185–91.

11. Rathod, P. K., McErlean, T., and Lee, P. C. 1997. Variations in frequencies of drug resistance in *Plasmodium falciparum. Proc. Natl. Acad. Sci. USA* 94:9389–93; Le Bras J., and Durand, R. 2003. The mechanisms of resistance to antimalarial drugs in *Plasmodium falciparum. Fundam. Clin. Pharmacol.* 17:147–53.

12. Gassis, S., and Rathod, P. K. 1996. Frequency of drug resistance in *Plasmodium falciparum:* a nonsynergistic combination of 5-fluoroorotate and atovaquone suppresses in vitro resistance. *Antimicrob. Agents Chemother.* 40:914–19.

13. White, N. J. 1999. Delaying antimalarial drug resistance with combination chemotherapy. *Parassitologia* 41:301–8.

14. Gassis and Rathod. 1996.

15. White. 1999.

16. White, N. J. 2004. Antimalarial drug resistance. *J. Clin. Invest.* 113:1084–92.

17. White. 1999.

18. Takahata, N. 1993. Allelic genealogy and human evolution. *Mol. Biol. Evol.* 10:2–22. Here I am considering the "census" population size—the actual

number of humans. The "effective" human population size—which is calculated from the genetic diversity of the population—is much smaller, perhaps only ten thousand or so.

19. Ten million years divided by one generation per ten years times a million creatures per generation.

20. The effective population size of the prolific house mouse has been estimated at between 450,000 and 810,000 (Keightley, P. D., Lercher, M. J., and Eyre-Walker, A. 2005. Evidence for widespread degradation of gene control regions in hominid genomes. *PLoS. Biol.* 3:e42). For the ancestor of humans and chimps the estimate is 20,000 (Rannala, B., and Yang, Z. 2003. Bayes estimation of species divergence times and ancestral population sizes using DNA sequences from multiple loci. *Genetics* 164:1645–56).

21. Mammals are thought to have arisen about 250 million years ago.

22. Le Bras, J., Durand, R. 2003. The mechanisms of resistance to antimalarial drugs in *Plasmodium falciparum*. *Fundam. Clin. Pharmacol.* 17:147–53.

23. Whitman, W. B., Coleman, D. C., Wiebe, W. J. 1998. Prokaryotes: the unseen majority. *Proc. Natl. Acad. Sci. USA* 95:6578–83.

4   What Darwinism Can Do

1. Mayr, E. 1991. *One long argument: Charles Darwin and the genesis of modern evolutionary thought.* Cambridge, Mass.: Harvard University Press, p. 36.

2. Darwin, C. 1859. *The origin of species.* New York: Bantam Books.

3. The genetic code is redundant, with some amino acids being coded for by multiple three-nucleotide sequences of DNA. So if a mutation merely switches one such redundant sequence for another, no change in amino acid sequence results.

4. National Library of Medicine. Genes and Disease. National Institutes of Health. 2004. The normal role of huntingtin is currently unknown.

5. Greenwood, B. 2002. The molecular epidemiology of malaria. *Trop. Med. Int. Health* 7:1012–21.

6. Lim, A. S., Cowman, A. F. 1996. *Plasmodium falciparum*: chloroquine selection of a cloned line and DNA rearrangements. *Exp. Parasitol.* 83:283–94.

7. Drake, J. W., Charlesworth, B., Charlesworth, D., and Crow, J. F. 1998. Rates of spontaneous mutation. *Genetics* 148:1667–86.

8. Lynch, M., Conery, J. S. 2000. The evolutionary fate and consequences of duplicate genes. *Science* 290:1151–55. However, other workers have estimated a much lower rate of gene duplication (Gao, L. Z., Innan, H. 2004. Very low gene duplication rate in the yeast genome. *Science* 306:1367–70).

9. Dayhoff, M. O., and National Biomedical Research Foundation. 1973.

Atlas of protein sequence and structure, vol. 5: supplement. Silver Spring, Md.: National Biomedical Research Foundation.

10. Chang, L. Y., and Slightom, J. L. 1984. Isolation and nucleotide sequence analysis of the beta-type globin pseudogene from human, gorilla and chimpanzee. *J. Mol. Biol.* 180:767–84.

11. Bapteste, E., Susko, E., Leigh, J., MacLeod, D., Charlebois, R. L., and Doolittle, W. F. 2005. Do orthologous gene phylogenies really support tree-thinking? *BMC Evol. Biol.* 5:33.

12. Dujon, B., et al. 2004. Genome evolution in yeasts. *Nature* 430:35–44.

13. Kellis, M., Birren, B. W., Lander, E. S. 2004. Proof and evolutionary analysis of ancient genome duplication in the yeast *Saccharomyces cerevisiae.* *Nature* 428:617–24.

14. Dujon et al. 2004.

15. Other genome duplications also seem to have had little discernible effect. For example, about half of all angiosperm (flowering plant) species are polyploid, and the number varies according to family and genus (Coyne, J. A., and Orr, H. A. 2004. *Speciation.* Sunderland, Mass.: Sinauer Associates, Chapter 9). Thus polyploidy apparently has no overwhelming advantage or disadvantage. The vertebrate lineage is thought to have undergone several rounds of genome duplication, which was speculated to have contributed to organismal complexity (Sidow, A. 1996. Gen(om)e duplications in the evolution of early vertebrates. *Curr. Opin. Genet. Dev.* 6:715–22). However, other workers see no connection (Donoghue, P. C. J., and Purnell, M. A. 2005. Genome duplication, extinction and vertebrate evolution. *Trends in Ecology and Evolution* 20:312–19).

16. Hayton, K. and Su, X. Z. 2004. Genetic and biochemical aspects of drug resistance in malaria parasites. *Curr. Drug Targets Infect. Disord.* 4:1–10.

17. Ibid.

18. Vaughan, A., Rocheleau, T., and ffrench-Constant, R. 1997. Site-directed mutagenesis of an acetylcholinesterase gene from the yellow fever mosquito *Aedes aegypti* confers insecticide insensitivity. *Exp. Parasitol.* 87:237–44.

19. "Non-silent point mutations within structural genes are the most common cause of target-site resistance. For selection of the mutations to occur, the resultant amino acid change must reduce the binding of the insecticide without causing a loss of primary function of the target site. Therefore the number of possible amino acid substitutions is very limited. Hence, identical resistance-associated mutations are commonly found across highly diverged taxa" (Hemingway, J., and Ranson, H. 2000. Insecticide resistance in insect vectors of human disease. *Annu. Rev. Entomol.* 45:371–91).

20. Pelz, H. J., Rost, S., Hunerberg, M., Fregin, A., Heiberg, A. C., Baert, K., Macnicoll, A. D., Prescott, C. V., Walker, A. S., Oldenburg, J., and Muller, C. R. 2005. The genetic basis of resistance to anticoagulants in rodents. *Genetics* 170:1839–47.

21. A similar mutation has also been observed in a human patient treated with the chemical as an anticoagulant.

22. Chen, I., DeVries, A. L., Cheng, C. H. 1997. Evolution of antifreeze glycoprotein gene from a trypsinogen gene in Antarctic notothenioid fish. *Proc. Natl. Acad. Sci. USA* 94:3811–16.

23. The sequence is not a perfect repeat. Apparently, some point mutations have also accumulated.

24. Chen et al. 1997.

25. Cheng, C. H., and Chen L. 1999. Evolution of an antifreeze glycoprotein. *Nature* 401:443–44.

26. Davies, P. L., Baardsnes, J., Kuiper, M. I., Walker, V. K. 2002. Structure and function of antifreeze proteins. *Philos. Trans. R. Soc. Lond. B. Biol. Sci.* 357:927–35.

27. Zachariassen, K. E., and Kristiansen, E. 2000. Ice nucleation and antinucleation in nature. *Cryobiology* 41:257–79.

28. Coluzzi, M. 1999. The clay feet of the malaria giant and its African roots: hypotheses and inferences about origin, spread and control of *Plasmodium falciparum*. *Parassitologia* 41:277–83; Sachs, J., and Malaney, P. 2002. The economic and social burden of malaria. *Nature* 415:680–85.

5   What Darwinism Can't Do

1. Kozminski, K. G., Johnson, K. A., Forscher, P., and Rosenbaum, J. L. 1993. A motility in the eukaryotic flagellum unrelated to flagellar beating. *Proc. Natl. Acad. Sci. USA* 90:5519–23; Sloboda, R. D. 2002. A healthy understanding of intraflagellar transport. *Cell Motil. Cytoskeleton* 52:1–8.

2. www.yale.edu/rosenbaum/rosen research.html.

3. Cole, D. G. 2003. The intraflagellar transport machinery of *Chlamydomonas reinhardtii*. *Traffic* 4:435–42.

4. Song, L., and Dentler, W. L. 2001. Flagellar protein dynamics in *Chlamydomonas*. *J. Biol. Chem.* 276:29754–63.

5. Cole, D. G. 2003. The intraflagellar transport machinery of *Chlamydomonas reinhardtii*. *Traffic* 4:435–42. When the amount of a protein that's normally part of complex A was experimentally decreased, the rest of the proteins found in complex A went down, too, but not those in complex B. So A and B might be separately regulated in the cell.

6. Apparently, IFT is needed to ensure that nodal cilia are made. Nodal cilia

are somewhat altered compared to most cilia. (They have a 9+0 structure rather than the common 9+2 structure, lacking the two central single microtubules of most cilia.) In the mutant mice nodal cilia were completely missing. Nodal cilia, which have a curious circular motion rather than the typical ciliary back-and-forth motion, are hypothesized to drive a flow of liquid that carries developmental factors to their proper positions (Scholey, J. M. 2003. Intraflagellar transport. *Annu. Rev. Cell Dev. Biol.* 19:423–43).

7. Pazour, G. J., Baker, S. A., Deane, J. A., Cole, D. G., Dickert, B. L., Rosenbaum, J. L., Witman, G. B., and Besharse, J. C. 2002. The intraflagellar transport protein, IFT88, is essential for vertebrate photoreceptor assembly and maintenance. *J. Cell Biol.* 157:103–13.

8. Marszalek, J. R., Liu, X., Roberts, E. A., Chui, D., Marth, J. D., Williams, D. S., and Goldstein, L. S. 2000. Genetic evidence for selective transport of opsin and arrestin by kinesin-II in mammalian photoreceptors. *Cell* 102:175–87.

9. Sloboda, R. D. 2002. A healthy understanding of intraflagellar transport. *Cell Motil. Cytoskeleton* 52:1–8.

10. Farley, J. 1977. *The spontaneous generation controversy from Descartes to Oparin*. Baltimore: Johns Hopkins University Press, p. 73.

11. The best example, in my opinion, is by University of Delaware biology professor John McDonald (http://udel.edu/~mcdonald/oldmousetrap.html).

12. Berriman and coworkers write of trypanosomes: "The proteins of the flagellar axoneme appeared to be extremely well conserved. With the exception of tektin, there are homologs in the three genomes for all previously identified structural components as well as a full complement of flagellar motors and both complex A and complex B of the intraflagellar transport system. . . . Thus, the 9+2 axoneme, which arose very early in eukaryotic evolution, appears to be constructed around a core set of proteins that are conserved in organisms possessing flagella and cilia" (Berriman, M., et al. 2005. The genome of the African trypanosome *Trypanosoma brucei*. *Science* 309:416–22).

13. But there is never a shortage of Darwinian conjecture. For an example, see Jekely, G., and Arendt, D. 2006. Evolution of intraflagellar transport from coated vesicles and autogenous origin of the eukaryotic cilium. *Bioessays* 28:191–98. Published in the "Hypotheses" section of the journal, the paper contains statements such as, "What was the initial advantage of membrane polarisation? One possibility is that the specialised membrane patch was advantageous for directional sensing," and, "It is also possible that compartmentalisation was favoured because of the nature of the receptor-mediated signalling pathways." In other words, the paper offers speculation.

14. I discuss the state of the science literature ten years after its original

publication in the Afterword of the tenth-anniversary edition of *Darwin's Black Box*.

15. Another example is a paper that takes a "Darwinian perspective" on explaining the evolution of the Type III secretory system. It quotes Charles Darwin in the *Origin of Species* remarking, "On the view of descent with modification, we may conclude that the existence of organs in a rudimentary, imperfect, and useless condition, . . . might even have been anticipated in accordance with the views here explained." This may support Darwin's idea of common descent, but does not speak to random mutation/natural selection (Pallen, M. J., Beatson, S. A., and Bailey, C. M. 2005. Bioinformatics, genomics and evolution of nonflagellar type-III secretion systems: a Darwinian perspective. *FEMS Microbil. Rev.* 29:201–29).

16. Kalir, S., McClure, J., Pabbaraju, K., Southward, C., Ronen, M., Leibler, S., Surette, M. G., and Alon, U. 2001. Ordering genes in a flagella pathway by analysis of expression kinetics from living bacteria. *Science* 292:2080–83.

17. Zaslaver, A., Mayo, A. E., Rosenberg, R., Bashkin, P., Sberro, H., Tsalyuk, M., Surette, M. G., and Alon, U. 2004. Just-in-time transcription program in metabolic pathways. *Nat. Genet.* 36:486–91.

18. Pearson, H. 2006. Genetics: what is a gene? *Nature* 441:398–401.

6 Benchmarks

1. Coyne, J. A., and Orr, H. A. 2004. *Speciation*. Sunderland, Mass.: Sinauer Associates, p. 136.

2. Like the great majority of evolutionary biologists, Orr and Coyne assume an evolutionary framework involving only unintelligent processes, and are dismissive of intelligent design. See Coyne, J. A. The case against intelligent design. *The New Republic*. August 22, 2005; Orr, H. A. Devolution. *The New Yorker.* May 30, 2005.

3. Adam, D. Give six monkeys a computer, and what do you get? Certainly not the Bard. *The Guardian*. 5-9-2003.

4. Smith, J. M. 1970. Natural selection and the concept of a protein space. *Nature* 225:563–64.

5. Orr, H. A. 2003. A minimum on the mean number of steps taken in adaptive walks. *J. Theor. Biol.* 220:241–47.

6. Drake, J. W., Charlesworth, B., Charlesworth, D., and Crow, J. F. 1998. Rates of spontaneous mutation. *Genetics* 148:1667–86.

7. Kimura, M. 1983. *The neutral theory of molecular evolution.* Cambridge: Cambridge University Press.

8. For example, see Gavrilets, S. 2004. *Fitness landscapes and the origin of species*. Princeton, N.J.: Princeton University Press.

9. One idea, proposed by Sewall Wright, to get around a rugged landscape is called the "shifting balance theory" (Wright, S. 1982. The shifting balance theory and macroevolution. *Annu. Rev. Genet.* 16:1–19), where a large population is subdivided into smaller, local ones, which can become less fit, and then perhaps begin to ascend another nearby, higher fitness peak. Wright's idea remains controversial.

10. Orr, H. A. 2003. A minimum on the mean number of steps taken in adaptive walks. *J. Theor. Biol.* 220:241–47.

11. "While thus employed, the heavy pewter lamp suspended in chains over his head, continually rocked with the motion of the ship, and for ever threw shifting gleams and shadows of lines upon his wrinkled brow, till it almost seemed that while he himself was marking out lines and courses on the wrinkled charts, some invisible pencil was also tracing lines and courses upon the deeply marked chart of his forehead."

12. Jacob, F. 1977. Evolution and tinkering. *Science* 196:1161–66. Jacob argued that an engineer made special parts for each new job, whereas a tinkerer reused old parts. Since parts of organisms appear to be related, Jacob thought that pointed to blind groping rather than foresight. Yet his argument addresses common descent, not random mutation. Whatever the merits of the analogy of tinker versus engineer in reusing parts, it simply doesn't address the question of whether guidance is needed to make a structure. Jacob didn't try to explain how complex structures could arise by chance, or by tinkering.

13. Behe, M. J. 1996. *Darwin's black box: the biochemical challenge to evolution.* New York: The Free Press, p. 39.

7 The Two-Binding-Sites Rule

1. Alberts, B. 1998. The cell as a collection of protein machines: preparing the next generation of molecular biologists. *Cell* 92:291–94.

2. Woodson, S. A. 2005. Biophysics: assembly line inspection. *Nature* 438:566–67.

3. Nooren, I. M., and Thornton, J. M. 2003. Diversity of protein-protein interactions. *EMBO J.* 22:3486–92.

4. Pauling, L. 1940. A theory of the structure and process of formation of antibodies. *Journal of the American Chemical Society* 62:2643–57. Pauling won the Nobel Prize for Chemistry in 1954 and for Peace in 1962.

5. Perelson, A. S., and Oster, G. F. 1979. Theoretical studies of clonal selection: minimal antibody repertoire size and reliability of self-non-self discrimination. *J. Theor. Biol.* 81:645–70; Segel, L. A., and Perelson, A. S. 1989. Shape space: an approach to the evaluation of cross-reactivity effects, stability and controllability in the immune system. *Immunol. Lett.* 22:91–99; De Boer, R. J.,

and Perelson, A. S. 1993. How diverse should the immune system be? *Proc. Biol. Sci.* 252:171–75; Smith, D. J., Forrest, S., Hightower, R. R., and Perelson, A. S. 1997. Deriving shape space parameters from immunological data. *J. Theor. Biol.* 189:141–50.

6. A properly working protein will bind to just one or a few specific partners. However, because the interiors of most folded proteins are oily, if a protein accidentally unfolds ("denatures"), it might stick to almost everything in sight. Nonspecific aggregation is almost always detrimental to a cell (Bucciantini, M., Giannoni, E., Chiti, F., Baroni, F., Formigli, L., Zurdo, J., Taddei, N., Ramponi, G., Dobson, C. M., and Stefani, M. 2002. Inherent toxicity of aggregates implies a common mechanism for protein misfolding diseases. *Nature* 416:507–11). Cells take great care to avoid it, with systems that dispose of misfolded proteins. So such nonspecific aggregation is not a model for how functional, specific, protein-protein interactions could develop in the cell.

7. Gavin, A. C., et al. 2002. Functional organization of the yeast proteome by systematic analysis of protein complexes. *Nature* 415:141–47; Giot, L., et al. 2003. A protein interaction map of *Drosophila* melanogaster. *Science* 302:1727–36; Li, S., et al. 2004. A map of the interactome network of the metazoan *C. elegans*. *Science* 303:540–43; Butland, G., et al. 2005. Interaction network containing conserved and essential protein complexes in *Escherichia coli*. *Nature* 433:531–37; LaCount, D. J., et al. 2005. A protein interaction network of the malaria parasite *Plasmodium falciparum*. *Nature* 438:103–7; Gavin, A. C., et al. 2006. Proteome survey reveals modularity of the yeast cell machinery. *Nature* 440:631–36.

8. The dissociation constant is on the order of micromolar (Nissim, A., Hoogenboom, H. R., Tomlinson, I. M., Flynn, G., Midgley, C., Lane, D., and Winter, G. 1994. Antibody fragments from a "single pot" phage display library as immunochemical reagents. *EMBO J.* 13:692–98).

9. Dissociation constant on the order of nanomolar (Griffiths, A. D., Williams, S. C., Hartley, O., Tomlinson, I. M., Waterhouse, P., Crosby, W. L., Kontermann, R. E., Jones, P. T., Low, N. M., Allison, T. J., and Winter, G. 1994. Isolation of high affinity human antibodies directly from large synthetic repertoires. *EMBO J.* 13:3245–60).

10. Other proteins besides antibodies have been used to demonstrate the same point. For example, rather than antibodies, a protein called knottin was used by Winter to generate a shape-space library. About five hundred million binding sites had to be screened to find one that stuck to a test protein with modest affinity (Smith, G. P., Patel, S. U., Windass, J. D., Thornton, J. M., Winter, G., and Griffiths, A. D. 1998. Small binding proteins selected from a combinatorial repertoire of knottins displayed on phage. *J. Mol. Biol.* 277:317–32).

A roughly similar result was seen when part of the sequence of a small bacterial protein was randomized to generate a library containing about forty million members, and when a library of fifty million artificial "minibodies" was probed for binding to a protein called interleukin (Nord, K., Gunneriusson, E., Uhlen, M., and Nygren, P. A. 2000. Ligands selected from combinatorial libraries of protein A for use in affinity capture of apolipoprotein A-1M and taq DNA polymerase. *J. Biotechnol.* 80:45–54; Martin, F., Toniatti, C., Salvati, A. L., Venturini, S., Ciliberto, G., Cortese, R., and Sollazzo, M. 1994. The affinity-selection of a minibody polypeptide inhibitor of human interleukin-6. *EMBO J.* 13:5303–9).

11. Most proteins are present in the cell at well below millimolar concentrations, so in order for two proteins to spend the majority of their time bound to each other, micromolar dissociation constants would be required to form even a "weak, transient" complex (Nooren, I. M., and Thornton, J. M. 2003. Structural characterisation and functional significance of transient protein-protein interactions. *J. Mol. Biol.* 325:991–1018). Dissociation constants on the order of micromolar seem to be required to detect interactions in yeast two-hybrid assays (Golemis, E. A., and Serebriiskii, I. 1996. Identification of protein-protein interactions. In Coligan, J. E., ed. *Current protocols in protein science.* Brooklyn, N.Y.: John Wiley & Sons, Inc; Estojak, J., Brent, R., and Golemis, E. A. 1995. Correlation of two-hybrid affinity data with in vitro measurements. *Mol. Cell Biol.* 15:5820–29).

12. As discussed in Chapter 7, there are different kinds of mutations—deletions, duplications, and so on. But point mutation represents the conceptually simplest, most straightforward route. This calculation uses consensus values for important variables. One could certainly imagine other scenarios for making a new protein-binding site, for example by first invoking gene duplication and then point mutation. But those are either unlikely to help much (Behe, M. J., and Snoke, D. W. 2004. Simulating evolution by gene duplication of protein features that require multiple amino acid residues. *Protein Sci.* 13:2651–64) or likely to involve special circumstances that amount to a Just-So story. All alternative scenarios would have to confront the fact that no new binding sites have turned up in the best-studied evolutionary cases of malaria and HIV, as described later in the text.

13. Even though protein-binding sites often involve a score of amino acids on each of the partners, experiments have shown that only a fraction of those are important for having the two proteins stick to each other. (For example, see Braden, B. C., and Poljak, R. J. 1995. Structural features of the reactions between antibodies and protein antigens. *FASEB J.* 9:9–16; Lo Conte, L., Chothia, C., and Janin, J. 1999. The atomic structure of protein-protein recognition sites. *J. Mol. Biol.* 285:2177–98; Ma, B., Elkayam, T., Wolfson, H., and

Nussinov, R. 2003. Protein-protein interactions: structurally conserved residues distinguish between binding sites and exposed protein surfaces. *Proc. Natl. Acad. Sci. USA* 100:5772–77.) In terms of the swimming pool analogy, the five or six residues represent bumps and magnets that are aligned very nicely; if enough are aligned, then it doesn't matter so much if other features aren't aligned, as long as they don't actively block the surfaces from coming together.

14. Axe, D. D. 2004. Estimating the prevalence of protein sequences adopting functional enzyme folds. *J. Mol. Biol.* 341:1295–1315.

15. Geretti, A. M. 2006. HIV-1 subtypes: epidemiology and significance for HIV management. *Curr. Opin. Infect. Dis.* 19:1–7. Rodrigo, A. G. 1999. HIV evolutionary genetics. *Proc. Natl. Acad. Sci. USA* 96:10559–61. Total body burden of the number of copies of HIV RNA is estimated to be much higher, about 10^{11} (Haase, A. T., Henry, K., Zupancic, M., Sedgewick, G., Faust, R. A., Melroe, H., Cavert, W., Gebhard, K., Staskus, K., Zhang, Z. O., Dailey, P. J., Balfour, H. H., Jr., Erice, A., and Perelson, A. S. 1996. Quantitative image analysis of HIV-1 infection in lymphoid tissue. *Science* 274:985–89). The effective population size is estimated at 500 to 10^5 (Althaus, C. L., and Bonhoeffer, S. 2005. Stochastic interplay between mutation and recombination during the acquisition of drug resistance mutations in human immunodeficiency virus type 1. *J. Virol.* 79:13572–78).

16. Rodrigo, A. G., Shpaer, E. G., Delwart, E. L., Iversen, A. K., Gallo, M. V., Brojatsch, J., Hirsch, M. S., Walker, B. D., and Mullins, J. I. 1999. Coalescent estimates of HIV-1 generation time in vivo. *Proc. Natl. Acad. Sci. USA* 96:2187–91.

17. Coffin, J. M. 1995. HIV population dynamics in vivo: implications for genetic variation, pathogenesis, and therapy. *Science* 267:483–89.

18. For example, see Viard, M., Parolini, I., Rawat, S. S., Fecchi, K., Sargiacomo, M., Puri, A., and Blumenthal, R. 2004. The role of glycosphingolipids in HIV signaling, entry and pathogenesis. *Glycocon. J.* 20:213–22; Bandivdekar, A. H., Velhal, S. M., and Raghavan, V. P. 2003. Identification of CD4-independent HIV receptors on spermatozoa. *Am. J. Reprod. Immunol.* 50:322–27. The virus also is taken up by dendritic cells, whose job is to transport them to lymphoid organs. For a short review, see Stebbing, J., Gazzard, B., and Douek, D. C. 2004. Where does HIV live? *N. Engl. J. Med.* 350:1872–80.

19. HIV, Human Immunodeficiency Virus, is thought to have originated from a similar virus infecting other primates, SIV, Simian Immunodeficiency Virus. Is the ability of the virus to change host species a major biochemical novelty? No. Apparently several changes in a small, variable region of one SIV protein were sufficient to allow SIV from sooty mangabeys to successfully infect Rhesus macaques (Demma, L. J., Logsdon, J. M., Vanderford, T. H.,

Feinberg, M. B., and Staprans, S. I. 2005. SIVsm quasispecies adaptation to a new simian host. *PLoS Pathol.* 1:e3).

20. What if something useful did happen, but so far has escaped detection? That of course can never be ruled out, but is unlikely. By definition, if something "useful" to the virus occurred, the mutant virus would increase rapidly in the population. Since the progression of HIV in the world is monitored closely, such an event would likely be noticed.

21. Wang (Wang, J. 2002. Protein recognition by cell surface receptors: physiological receptors versus virus interactions. *Trends Biochem. Sci.* 27:122–26) argues that viruses must bind their receptors more strongly than the normal physiological receptor ligand. However, he points out that normal ligands don't bind very strongly on an individual basis, relying on the avidity of multivalent binding to compensate. So a viral protein that bound a cell surface protein with a modest dissociation constant of one micromolar or better should be sufficient. Micromolar dissociation constants would be expected to be found in a shape-space library of 10^8 surfaces. That many mutant viruses would be present in a single individual each day. However, the mutations would not likely be clustered in a coherent patch, as they are in the shape-space libraries that Winter developed.

22. Sarafianos, S. G., Das, K., Hughes, S. H., and Arnold, E. 2004. Taking aim at a moving target: designing drugs to inhibit drug-resistant HIV-1 reverse transcriptases. *Curr. Opin. Struct. Biol.* 14:716–30.

23. Tie, Y., Boross, P. I., Wang, Y. F., Gaddis, L., Liu, F., Chen, X., Tozser, J., Harrison, R. W., and Weber, I. T. 2005. Molecular basis for substrate recognition and drug resistance from 1.1 to 1.6 angstroms resolution crystal structures of HIV-1 protease mutants with substrate analogs. *FEBS J.* 272:5265–77.

24. The evasion of the immune system by HIV is also biochemically simple, involving point mutations in the outer proteins of the virus coat, which apparently causes antibodies that had been produced against the virus to bind less strongly. For example, see Frost, S. D., Wrin, T., Smith, D. M., Pond, S. L., Liu, Y., Paxinos, E., Chappey, C., Galovich, J., Beauchaine, J., Petropoulos, C. J., Little, S. J., and Richman, D. D. 2005. Neutralizing antibody responses drive the evolution of human immunodeficiency virus type 1 envelope during recent HIV infection. *Proc. Natl. Acad. Sci. USA* 102:18514–19.

25. Lenski, R. E. 2004. Phenotypic and genomic evolution during a 20,000-generation experiment with the bacterium *Escherichia coli*. *Plant Breeding Reviews* 24:225–65; Cooper, T. F., Rozen, D. E., and Lenski, R. E. 2003. Parallel changes in gene expression after 20,000 generations of evolution in *Escherichia coli*. *Proc. Natl. Acad. Sci. USA* 100:1072–77; Cooper, V. S., Schneider, D., Blot, M., and Lenski, R. E. 2001. Mechanisms causing rapid and parallel losses of ribose catabolism in evolving populations of *Escherichia coli*

B. *J. Bacteriol.* 183:2834–41; Vulic, M., Lenski, R. E., and Radman, M. 1999. Mutation, recombination, and incipient speciation of bacteria in the laboratory. *Proc. Natl. Acad. Sci. USA* 96:7348–51.

26. "Experimental populations of *Escherichia coli* have evolved for 20,000 generations in a uniform environment. Their rate of improvement, as measured in competitions with the ancestor in that environment, has declined substantially over this period. . . . Instead, the pronounced deceleration in its rate of fitness improvement indicates that the population early on incorporated most of those mutations that provided the greatest gains, and subsequently relied on beneficial mutations that were fewer in number, smaller in effect, or both" (de Visser, J. A., and Lenski, R. E. 2002. Long-term experimental evolution in *Escherichia coli*. XI. Rejection of non-transitive interactions as cause of declining rate of adaptation. *BMC Evol. Biol.* 2:19).

8 Objections to the Edge

1. Proteins that interact with chemicals, especially small molecules in metabolism, also constitute a separate category. One reason for placing these in a separate category is that metabolites are often present at much higher concentrations in the cell than are proteins.

2. For example, some viral proteins stick to cell proteins that normally trigger defensive action, either by alerting the immune system that the cell has been invaded or by tripping a mechanism that makes compromised cells self-destruct (Tortorella, D., Gewurz, B. E., Furman, M. H., Schust, D. J., and Ploegh, H. L. 2000. Viral subversion of the immune system. *Annu. Rev. Immunol.* 18:861–926; Gale, M., Jr., and Foy, E. M. 2005. Evasion of intracellular host defence by hepatitis C virus. *Nature* 436:939–45).

3. Deoxy HbS has a solubility of about 3 millimolar (Behe, M. J., and Englander, S. W. 1979. Mixed gelation theory. Kinetics, equilibrium and gel incorporation in sickle hemoglobin mixtures. *J. Mol. Biol.* 133:137–60). It's dissociation constant, however, is much higher due to excluded volume effects (Ivanova, M., Jasuja, R., Kwong, S., Briehl, R. W., and Ferrone, F. A. 2000. Nonideality and the nucleation of sickle hemoglobin. *Biophys. J.* 79:1016–22).

4. Rollins, C. T., Rivera, V. M., Woolfson, D. N., Keenan, T., Hatada, M., Adams, S. E., Andrade, L. J., Yaeger, D., van Schravendijk, M. R., Holt, D. A., Gilman, M., and Clackson, T. 2000. A ligand-reversible dimerization system for controlling protein-protein interactions. *Proc. Natl. Acad. Sci. USA* 97:7096–7101.

5. Stronger binding isn't necessarily better for a protein. One protein called CA that coats the RNA genome of HIV has amino acids that decrease the bind-

ing of the protein to itself, apparently allowing it to do its job better (del Alamo, M., Neira, J. L., and Mateu, M. G. 2003. Thermodynamic dissection of a low affinity protein-protein interface involved in human immunodeficiency virus assembly. *J. Biol. Chem.* 278:27923–29).

6. Proteins sometimes take on the altered shapes of other proteins in neurological diseases such as Alzheimer's (Dobson, C., M. 2002. Protein-misfolding diseases: getting out of shape. *Nature* 418:729–30). Also, some proteins that bind copies of themselves are thought to have arisen by "domain swapping." That is, areas called "domains" that had been in contact in a single protein instead bind to the analogous region of a second copy of the protein. This scenario does not propose that new binding sites arose, just that pre-existing binding sites got mixed up (Rousseau, F., Schymkowitz, J. W., and Itzhaki, L., S. 2003. The unfolding story of three-dimensional domain swapping. *Structure* 11:243–51).

7. Actually for many organisms the mutation rate per cell duplication is even lower, more like one in 10^{10}. However, for animals like humans, several hundred generations of cells pass between the birth of a parent and the birth of a child, so the rate of inherited nucleotide mutations per generation is about one in 10^8 (Drake, J. W., Charlesworth, B., Charlesworth, D., and Crow, J. F. 1998. Rates of spontaneous mutation. *Genetics* 148:1667–86).

8. This assumes a genome of 10^8 nucleotides or larger.

9. Voet, D., and Voet, J. G. 2004. *Biochemistry*. New York: J. Wiley & Sons, p. 14.

10. Kauffman, S. A. 1995. *At home in the universe: the search for laws of self-organization and complexity*. New York: Oxford University Press. Camazine (2001) gives many persuasive examples of self-organizing *behavior* in biological systems (almost entirely at the organismal, rather than molecular, level) but gives no examples of the self-organization of *evolution* (Camazine, S., et al. 2001. *Self-organization in biological systems*. Princeton, N.J.: Princeton University Press).

11. Shapiro, J. A., and Sternberg R. V. 2005. Why repetitive DNA is essential to genome function. *Biological Reviews* 80:227–50.

12. Some scientists such as Shapiro restrict the term "Darwinian" to mutations of small effect, and thus think of big events such as gene duplications or transpositions as "non-Darwinian."

13. Han, J. S., Szak, S. T., and Boeke, J. D. 2004. Transcriptional disruption by the L1 retrotransposon and implications for mammalian transcriptomes. *Nature* 429:268–74.

14. Fondon, J. W., III, and Garner, H. R. 2004. Molecular origins of rapid and continuous morphological evolution. *Proc. Natl. Acad. Sci. USA* 101:18058–63.

15. Ronshaugen, M., McGinnis, N., and McGinnis, W. 2002. Hox protein mutation and macroevolution of the insect body plan. *Nature* 415:914–17.

16. Schneider, D., Duperchy, E., Coursange, E., Lenski, R. E., and Blot, M. 2000. Long-term experimental evolution in *Escherichia coli*. IX. Characterization of insertion sequence-mediated mutations and rearrangements. *Genetics* 156:477–88.

17. Maxwell, J. C. 1952. *The scientific papers of James Clerk Maxwell*. New York: Dover Publications; "Ether," *Encyclopaedia Brittanica*, ninth edition, pp. 763–75.

18. One objection to the shipwrecked-sailor analogy might be the following: We know humans are intelligent agents, so we can conclude that a sailor might have arranged some features of the island, but we can't conclude design if we don't have a human designer in the wings. To understand why that reasoning is wrong, instead of a deserted island, just change the example to a tellurian spaceship that crashes on an unexplored planet. When the marooned astronaut spots a wrecked, alien ship over a hill, the same reasoning applies, now with a nonhuman intelligent agent.

19. "The purposeful or inventive arrangement of parts or details," www.thefreedictionary.com/design. In *Darwin's Black Box* (p. 193), I defined design as "the purposeful arrangement of parts."

20. For example, suppose the shipwrecked sailor arranged rocks throughout the island to point to various constellations in the night sky. You might easily not notice the arrangement, even if you saw the individual rocks.

9   The Cathedral and the Spandrels

1. Kirschner, M., and Gerhart, J. 2005. *The plausibility of life: resolving Darwin's dilemma*. New Haven, Conn.: Yale University Press, p. 53.

2. Laughon, A., and Scott, M. P. 1984. Sequence of a *Drosophila* segmentation gene: protein structure homology with DNA-binding proteins. *Nature* 310:25–31. Subsequent work has shown that homeodomain helix-turn-helix motifs are somewhat different from bacterial ones (Branden, C., and Tooze, J. 1999. *Introduction to protein structure*, 2nd ed. New York: Garland Publishing /Taylor and Francis).

3. Davidson, E. H. 2000. *Genomic regulatory systems: development and evolution*. San Diego: Academic Press, pp. 11–12.

4. A good introductory description of *Drosophila* development can usually be found in any recent textbook of molecular biology or developmental biology. For example, Alberts, B. 2002. *Molecular biology of the cell*, 4th ed. New York: Garland Science; Gilbert, S. F., Singer, S. R., Tyler, M. S., and Kozlowski,

R. N. 2003. *Developmental biology,* 7th ed. Sunderland, Mass.: Sinauer Associates.

5. The messenger RNA for the proteins is deposited.

6. Gehring, W. J. 1996. The master control gene for morphogenesis and evolution of the eye. *Genes Cells* 1:11–15.

7. Wagner, G. P., and Schlosser, G. 2004. *Modularity in development and evolution.* Chicago: University of Chicago Press; Callebaut, W., and Rasskin-Gutman, D. 2005. *Modularity: understanding the development and evolution of natural complex systems.* Cambridge, Mass.: MIT Press.

8. Callebaut, W., and Rasskin-Gutman, D. 2005, sections III and IV.

9. There are four different kinds of nucleotides in DNA (A, C, G, and T). So the probability of matching a six-nucleotide-long binding site is one in four to the sixth power, which is one in 4,096.

10. Stone, J. R., Wray, G. A. 2001. Rapid evolution of *cis*-regulatory sequences via local point mutations. *Mol. Biol. Evol.* 18:1764–70.

11. A further study confirmed the notion that regulatory protein switch sequences are a dime a dozen: Almost one in a hundred of all random eighteenth-nucleotide-long DNA fragments had the ability to increase the rate at which a test gene was expressed. (Edelman, G. M., Meech, R., Owens, G. C., and Jones, F. S. 2000. Synthetic promoter elements obtained by nucleotide sequence variation and selection for activity. *Proc. Natl. Acad. Sci. USA* 97:3038–43).

12. Stone, J. R., and Wray, G. A. 2001. The following quotation omits references cited in the paper.

It is likely that in many cases, the appearance of a new binding site will have no effect on the expression of a nearby gene. For example, some transcription factors must interact with other DNA-binding proteins to affect gene expression, and therefore the appearance of a single new binding site might be functionally neutral. Alternatively, nearby silencer sites might override any new binding site that appears. Nevertheless, it is plausible that in some cases the appearance of a new binding site will alter gene expression. Because most promoters are inactive by default, activation of transcription by the appearance of a single new binding site seems less likely than does modulation or restriction of an existing phase of gene expression. Of course, even if a new binding site does affect gene expression, there is no way to predict what, if any, phenotypic consequences might ensue. As with amino acid substitutions within coding regions of genes, we predict that in many cases the consequences of a new binding site appearing within a promoter will be either detrimental or neutral; only in rare cases will it be beneficial. The salient prediction yielded by our computer simulation is that local point mutations will constantly produce new binding sites that *in principle* [emphasis added] are capable of altering gene expression and that such sites may be subject to selection.

It is useful to reiterate here the Coyne-Orr principle, that "the goal of theory, however, is to determine not just whether a phenomenon is theoretically possible, but whether it is *biologically reasonable*—that is, whether it occurs with significant frequency under conditions that are likely to occur in nature." In other words, showing something could occur "in principle" should not be the end of the story.

13. Stone, J. R., and Wray, G. A. 2001.

14. Coyne, J. A. 2005. Switching on evolution. *Nature* 435:1029–30.

15. "The origin of insect wings has long been a contentious mystery. . . . But here again is where Evo Devo has stepped in with some powerful new evidence. . . . In order to test the theory that wings might be derived from the gill branches of crustaceans, Michalis Averof and Stephen Cohen traced how the Apterous and Nubbin [master regulatory] proteins are expressed in the appendage of other arthropods, especially crustaceans. They found [the proteins] were selectively expressed in the respiratory lobe of the outer branch of crustacean limbs. The best explanation for this observation is that the respiratory lobe and insect wing are homologous—that is, the same body part in different forms in the two animals" (Carroll, S. B. 2005. *Endless forms most beautiful: the new science of evo devo and the making of the animal kingdom.* New York: W.W. Norton & Co., pp. 175–76).

16. Carroll dismisses another eventuality. "The only other possibility would be the extraordinary coincidence that, of the hundreds of tool kit proteins that could be used to make gills and wings, these two proteins were independently selected by crustaceans and insects for building these structures." I find this dismissal odd from someone who touts the extraordinary power of switches and regulatory proteins. If those two proteins could trigger some useful developmental pathway, and if, like eyes on antennae, the pathway could be triggered at some arbitrary point in the animal's body, why should an evo-devo devotee not think wings are a new invention?

17. Erwin, D. H. 2005. A variable look at evolution. *Cell* 123:177–79; Charlesworth, B. 2005. On the origins of novelty and variation. *Science* 310:1619–20; Hartl, D. L. 2005. Better living through evolution. *Harvard Magazine* (November–December), pp. 22–27.

18. quoted in Hartl, 2005. Hartl tries to give this a positive spin. "Darwin

could not have dreamed of such a spectacular confirmation of his theory of descent with modification [that is, similar molecules in different animals]." But Jacob was using Darwin's theory to reason that cows should have cow molecules. If that had transpired, no Darwinist would have been surprised. The surprise to Darwinists was that their expectations based on Darwin's theory were wrong, that cows in fact didn't have special cow molecules. Hartl goes on: "For creationists, this must be a nightmare, for any sensible model of creationism would predict cows to have cow molecules, goats to have goat molecules, and snakes to have snake molecules." But it was Jacob and other leading Darwinists themselves who confidently expected different molecules for different species.

19. Carroll. 2005, pp. 71–72.

20. Ibid., pp. 123, 318.

21. Ibid., p. 285.

22. Ibid., p. 173.

23. Kirschner, M., and Gerhart, J. 2005, p. 195.

24. Ibid., p. 196.

25. Gehring. 1996.

26. Alberts, B. 1998. The cell as a collection of protein machines: preparing the next generation of molecular biologists. *Cell* 92:291–94.

27. Erwin, D. H. 2005. The books Erwin cites are: Kirschner, M., and Gerhart, J. 2005. *The plausibility of life: great leaps of evolution*. New Haven, Conn.: Yale University Press; Carroll, S. B. 2005: *Endless forms most beautiful: the new science of evo devo and the making of the animal kingdom*. New York: W.W. Norton & Co; Jablonka, E., and Lamb, M. J. 2005. *Evolution in four dimensions: genetic, epigenetic, behavioral, and symbolic variation in the history of life*. Cambridge, Mass.: MIT Press; West-Eberhard, M. J. 2003. *Developmental plasticity and evolution*. Oxford: Oxford University Press; Schlichting, C., and Pigliucci, M. 1998. *Phenotypic evolution: a reaction norm perspective*. Sunderland, Mass.: Sinauer.

28. Kirschner, M., and Gerhart, J. 2005, p. 242.

29. Watson, R. A. 2006. *Compositional evolution: the impact of sex, symbiosis, and modularity on the gradualist framework of evolution*. Cambridge, Mass.: MIT Press, p. 272.

30. Carroll. 2005, p. 139.

31. The number of the tiniest animals such as nematode worms at best may be comparable to malaria, but larger animals such as insects are much fewer in number.

32. Orr, H. A. 2003. A minimum on the mean number of steps taken in adaptive walks. *J. Theor. Biol.* 220:241–47.

33. Stone, J. R., and Wray, G. A. 2001.

34. Levine, M., and Davidson, E. H. 2005. Gene regulatory networks for development. *Proc. Natl. Acad. Sci. USA* 102:4936–42.

35. Davidson, E. H., and Erwin, D. H. 2006. Gene regulatory networks and the evolution of animal body plans. *Science* 311:796–800.

36. Singh, H., Medina, K. L., and Pongubala, J. M. 2005. Contingent gene regulatory networks and B cell fate specification. *Proc. Natl. Acad. Sci. USA* 102:4949–53. See Figure 9. 3.

37. Valentine, J. W., Collins, A. G., and Meyer, C. P. 1994. Morphological complexity increase in metazoans. *Paleobiology* 20:131–42.

38. Davidson, E. H., pp. 11–12.

39. Fondon, J. W., III, and Garner, H. R. 2004. Molecular origins of rapid and continuous morphological evolution. *Proc. Natl. Acad. Sci. USA* 101:18058–63.

40. Culotta, E., and Pennisi, E. 2005. Breakthrough of the year: evolution in action. *Science* 310:1878–79.

41. Colosimo, P. F., Hosemann, K. E., Balabhadra, S., Villarreal, G., Jr., Dickson, M., Grimwood, J., Schmutz, J., Myers, R. M., Schluter, D., and Kingsley, D. M. 2005. Widespread parallel evolution in sticklebacks by repeated fixation of Ectodysplasin alleles. *Science* 307:1928–33.

42. Gompel, N., Prud'homme, B., Wittkopp, P. J., Kassner, V. A., and Carroll, S. B. 2005. Chance caught on the wing: *cis*-regulatory evolution and the origin of pigment patterns in *Drosophila*. *Nature* 433:481–87.

43. Gould, S. J., and Lewontin, R. C. 1979. The spandrels of San Marco and the Panglossian paradigm: a critique of the adaptationist programme. *Proc. R. Soc. Lond. B. Biol. Sci.* 205:581–98.

44. Gould, S. J. 1997. The exaptive excellence of spandrels as a term and prototype. *Proc. Natl. Acad. Sci. USA* 94:10750–55.

10   All the World's a Stage

1. White, M., and Gribbin, J. R. 2002. *Stephen Hawking: a life in science*, 2nd ed. Washington, D.C.: Joseph Henry Press, p. 261.

2. Carter, B. 1974. Large number coincidences and the anthropic principle in cosmology. In *Confrontation of cosmological theories with data*, M. S. Longair, ed. Dordrecht: Reidel, pp. 291–98. A recent, balanced discussion of fine-tuning is Collins, R. 2003. Evidence for fine-tuning. In *God and design: the teleological argument and modern science*. Neil Manson, ed. Routledge, pp. 178–99.

3. Davies, P. C. W. 1982. *The accidental universe*. Cambridge: Cambridge University Press, p. vii.

4. Hawking, S. W. 1998. *A brief history of time,* tenth anniversary ed. New York: Bantam Books, pp. 129–30.

5. Dyson, F. J. 1979. *Disturbing the universe.* New York: Harper & Row, p. 250.

6. Denton, M. J. 1998. *Nature's destiny: how the laws of biology reveal purpose in the universe.* New York: Free Press.

7. A possible objection is that chemical properties are derived from physical ones, so why count them as a separate category? As a matter of bookkeeping, I'm categorizing fine-tuning by the level at which the effects become noticeable to us. The fine-tuning of laws and constants becomes apparent at the level of physics and astronomy, such as whether atoms or stars would be stable. The properties of water and carbon that are necessary for life, however, become apparent only at the level of chemistry.

8. National Academy of Sciences. 1999. *Science and creationism: a view from the National Academy of Sciences,* 2nd ed. Washington, D. C.: National Academy Press, p. 7.

9. Ward, P. D., and Brownlee, D. 2000. *Rare earth: why complex life is uncommon in the universe.* New York: Copernicus, p. 29.

10. Ibid., p. 231, figure legend 10.3.

11. Davies, P. C. W. 1999. *The fifth miracle: The search for the origin and meaning of life.* New York: Simon & Schuster, p. 17.

12. Crick, F. 1981. *Life itself: its origin and nature.* New York: Simon and Schuster, p. 88.

13. Farley, J. 1977. *The spontaneous generation controversy from Descartes to Oparin.* Baltimore: Johns Hopkins University Press, p. 73.

14. Bostrom, N. 2002. *Anthropic bias: observation selection effects in science and philosophy.* New York: Routledge, p.12.

15. Brumfiel, G. 2006. Our universe: outrageous fortune. *Nature* 439:10–12; Carroll, S. M. 2006. Is our universe natural? *Nature* 440:1132–36.

16. Collins, R. 2002. The argument from design and the many-worlds hypothesis. In *Philosophy of religion: a reader and guide,* W. L. Craig, ed. New Brunswick, N. J.: Rutgers University Press.

17. It's entirely possible a designer might set up a multiverse in order to generate one or more life-containing universes. But of course that would not be random; it would be intentional. In this section I'm considering what to expect just from a random collection of unintended universes.

18. Assuming each person bought just one ticket, roughly one in a million.

19. On the other hand, even if they aren't absolutely necessary to produce intelligent life, a designer might make complex systems such as the flagellum, either for their own sake or for other, nonessential interactions with intelligent life.

20. Bostrom, N. 2002. *Anthropic bias: observation selection effects in science and philosophy.* New York: Routledge, pp. 52–53,55.

21. In one episode of the original *Star Trek* TV series ("Mirror, Mirror," number 39), Captain Kirk enters a parallel universe where he meets a rather sinister, bearded Mr. Spock, plus other unsavory doubles of the nice-in-our-universe crew.

22. If there are infinite universes, how can there be no combinations of laws and constants that would allow an orderly progression to life? Here's one possibility. One can have an infinite number of something, but still not have all values. For example, suppose in the multiverse some particular constant could take any real-numbered value between one and two. There are an infinite number of possible values within that range. But if life could arise only if the constant had a value of eight-and-a-half—a number outside the permitted range—then none of the infinite number of constants would work. Here's another possibility. Suppose that in reality there are no laws and constants of nature, only chaos. The chaos might by chance give rise to delusional freak observers, but not to observers who have arisen by laws of nature, since none exist.

23. Bostrom, N. 2002. *Anthropic bias: observation selection effects in science and philosophy.* New York: Routledge, pp. 11–12. Bostrom takes designer in a very broad sense—in fact, too broad. I believe his inclusion of the words "principle or mechanism" in the second sentence is a category mistake on Bostrom's part. A design hypothesis implies intentionality, choice, and other characteristics that cannot be attributed to a "principle or mechanism."

24. Sawyer, R. J. 2000. The abdication of Pope Mary III . . . or Galileo's revenge. *Nature* 406:23. The journal *Nature* is generally dismissive of scientific results suggestive of a reality beyond nature, whether the results be fine-tuning of the cosmos or a beginning to the universe. In 1989 the longtime editor of *Nature,* John Maddox wrote an editorial with the odd title "Down with the Big Bang." Maddox decried the Big Bang theory as "philosophically unacceptable," saying it gave aid and comfort to "Creationists."

25. Quoted in Kirschner, M., and Gerhart, J. 2005. *The plausibility of life: great leaps of evolution.* New Haven, Conn.: Yale University Press. p. 265.

26. Hall, B. G. 2004. In vitro evolution predicts that the IMP-1 metallo-beta-lactamase does not have the potential to evolve increased activity against imipenem. *Antimicrob. Agents Chemother.* 48:1032–33.

Appendix A I, Nanobot

1. A self-replicating robot was recently reported in *Nature* (Zykov, V., Mytilinaios, E., Adams, B., and Lipson, H. 2005. Robotics: self-reproducing machines. *Nature* 435:163–64).

2. Giles, J. 2004. Nanotech takes small step towards burying "grey goo." *Nature* 429:591.

3. The terms "robot" and "machine" applied to the cell are *not* meant as analogies—they are meant quite literally. That cells and the systems they contain are robotic machinery is widely recognized in the scientific community. For example, Tanford and Reynolds dub proteins "Nature's Robots" (Tanford, C., and Reynolds, J. A. 2001. *Nature's robots: a history of proteins*. Oxford: Oxford University Press) and the term "molecular machines" is routinely used to describe protein complexes. For example, see the December 2003 *BioEssays* Special Issue on Molecular Machines, containing such articles as "The spliceosome: the most complex macromolecular machine in the cell?" and "Perpetuating the double helix: molecular machines at eukaryotic DNA replication origins."

4. Much of the following discussion of the history of biology derives from Singer, C. J. 1959. *A history of biology to about the year 1900: a general introduction to the study of living things*, 3rd ed. London: Abelard-Schuman.

5. Singer. 1959.

6. Galen was born in Greece but traveled to Rome at the age of about thirty-two where he became the physician to the emperor.

7. The Englishman William Harvey first reasoned that blood had to circulate. He calculated that if each heartbeat pumped two ounces of blood, and if a heart beat seventy-two times per minute, then in an hour the heart pumps 540 pounds of blood—triple the weight of a large man! Clearly that much blood could not be continuously made by the body. Harvey's elegant mathematical reasoning is cited as one of the earliest examples of modern scientific thinking (Singer. 1959).

8. Joseph Hook called them "cells" because they reminded him of medieval monks' rooms.

9. So said Ernst Haeckel (Farley, J. 1977. *The spontaneous generation controversy from Descartes to Oparin*. Baltimore: Johns Hopkins University Press, p. 73).

10. Improvements in computers and lab techniques have made crystallography much faster and more tractable, although it still involves considerable effort (Abbott, A. 2005. Protein structures hint at the shape of things to come. *Nature* 435:547).

11. Perutz, M. 1964. The hemoglobin molecule. *Scientific American* 211:64–76.

12. Other classes of biological molecules include DNA and RNA (which carry genetic information), and polysaccharides and lipids (which have structural and energy-storage roles).

13. Discussions of protein structure, myoglobin, and hemoglobin can be found in virtually any biochemistry textbook.

14. Most proteins are less than a thousand amino acids long. A single DNA, however, can be composed of a hundred million nucleotides. That vast length of DNA can contain many discrete genes that code for many different proteins.

15. Chains of amino acids (proteins) that can fold into discrete and functional forms are exceedingly rare, although the exact degree of rarity is a matter of debate. Estimates range roughly from one sequence in every 10^{30} (Hecht, M. H., Das, A., Go, A., Bradley, L. H., and Wei, Y. 2004. De novo proteins from designed combinatorial libraries. *Protein Sci.* 13:1711–23)to about one in 10^{70} (Axe, D. D. 2004. Estimating the prevalence of protein sequences adopting functional enzyme folds. *J. Mol. Biol.* 341:1295–1315).

16. For conceptual clarity, in the "water-loving" category here I have included only uncharged hydrophilic amino acids, and put charged residues in their own group. However, charged amino acids are "water-loving," too. Some amino acids don't clearly fit any of the categories.

17. Myoglobin was long thought to be required to store oxygen in tissues of all mammals, but its role is no longer so clear. A few years ago researchers destroyed the gene for myoglobin in a line of mice, so that the mice were entirely missing myoglobin, and yet adult mice did just fine without it (Garry, D. J., Ordway, G. A., Lorenz, J. N., Radford, N. B., Chin, E. R., Grange, R. W., Bassel-Duby, R., and Williams, R. S. 1998. Mice without myoglobin. *Nature* 395:905–8). Later work showed that unborn mice do have some need for the protein. The point, however, is that the role myoglobin plays in animals, which was thought to be very well understood, was not (Kanatous, S. B., and Garry, D. J. 2006. Gene deletional strategies reveal novel physiological roles for myoglobin in striated muscle. *Respir. Physiol Neurobiol.* 151:151–58; Garry, D. J., Kanatous, S. B., and Mammen, P. P. 2003. Emerging roles for myoglobin in the heart. *Trends Cardiovasc. Med.* 13:111–16; Meeson, A. P., Radford, N., Shelton, J. M., Mammen, P. P., DiMaio, J. M., Hutcheson, K., Kong, Y., Elterman, J., Williams, R. S., and Garry, D. J. 2001. Adaptive mechanisms that preserve cardiac function in mice without myoglobin. *Circ. Res.* 88:713–20).

Appendix B   Malaria Drug Resistance

1. Sibley, C. H., Hyde, I. E., Sims, P. F., Plowe, C. V., Kublin, J. G., Mberu, E. K., Cowman, A. F., Winstanley, P. A., Watkins, W. M., and Nzila, A. M. 2001. Pyrimethamine-sulfadoxine resistance in *Plasmodium falciparum*: what next? *Trends Parasitol.* 17:582–88; Le Bras, J., and Durand, R. 2003. The mechanisms of resistance to antimalarial drugs in *Plasmodium falciparum. Fundam. Clin. Pharmacol.* 17:147–53. The authors point out that S/P should not be considered a drug combination because the sulfadoxine and pyrimethamine must

act synergistically. If resistance to either drug develops, the therapy fails. Thus only one mutation is required for resistance.

2. Plowe, C. V. 2003. Monitoring antimalarial drug resistance: making the most of the tools at hand. *J. Exp. Biol.* 206:3745–52; Kublin, J. G., Cortese, J. F., Njunju, E. M., Mukadam, R. A., Wirima, J. J., Kazembe, P. N., Djimde, A. A., Kouriba, B., Taylor, T. E., and Plowe, C. V. 2003. Reemergence of chloroquine-sensitive *Plasmodium falciparum* malaria after cessation of chloroquine use in Malawi. *J. Infect. Dis.* 187:1870–75.

3. Korsinczky, M., Chen, N., Kotecka, B., Saul, A., Rieckmann, K., and Cheng, Q. 2000. Mutations in *Plasmodium falciparum,* cytochrome *b* that are associated with atovaquone resistance are located at a putative drug-binding site. *Antimicrob. Agents Chemother.* 44:2100–2108.

4. Meshnick, S. R. 2002. Artemisinin: mechanisms of action, resistance and toxicity. *Int. J. Parasitol.* 32:1655–60; Uhlemann, A. C., Cameron, A., Eck-stein-Ludwig, U., Fischbarg, J., Iserovich, P., Zuniga, F. A., East, M., Lee, A., Brady, L., Haynes, R. K., and Krishna, S. 2005. A single amino acid residue can determine the sensitivity of SERCAs to artemisinins. *Nat. Struct. Mol. Biol.* 12:628–29.

5. White, N. J. 2004. Antimalarial drug resistance. *J. Clin. Invest.* 113:1084–92.

6. Ferdig, M. T., Cooper, R. A., Mu, J., Deng, B., Joy, D. A., Su, X. Z., Wellems, T. E., 2004. Dissecting the loci of low-level quinine resistance in malaria parasites. *Mol. Microbio.* 52:985–97; Mu, J., Ferdig, M. T., Feng, X., Joy, D. A., Duan, J., Furuya, T., Subramanian, G., Aravind, L., Cooper, R. A., Wootton, J. C., Xiong, M., Su, X. Z. 2003. Multiple transporters associated with malaria parasite responses to chloroquine and quinine. *Mol. Microbiol.* 49:977–89.

Appendix C   Assembling the Bacterial Flagellum

1. Voet, D., and Voet, J. G. 1995. *Biochemistry,* 2nd ed. New York: J. Wiley & Sons, p. 1260.

2. Much of the following description of the structure and assembly of the flagellum is based on Minamino, T., and Namba, K. 2004. Self-assembly and type III protein export of the bacterial flagellum. *J. Mol. Microbiol. Biotechnol.* 7:5–17.

3. Minamino and Namba, 2004.

4. www.npn.jst.go.jp.

5. See the video at www.npn.jst.go.jp.

6. Macnab, R. M. 1999. The bacterial flagellum: reversible rotary propellor and Type III export apparatus. *J. Bacteriol.* 181:7149–53; Nguyen, L., Paulsen,

I. T., Tchieu, J., Hueck, C. J., and Saier, M. H., Jr. 2000. Phylogenetic analyses of the constituents of Type III protein secretion systems. *J. Mol. Microbiol. Biotechnol.* 2:125–44; Galan, J. E., and Collmer, A. 1999. Type III secretion machines: bacterial devices for protein delivery into host cells. *Science* 284:1322–28; Saier, M. H., Jr. 2004. Evolution of bacterial Type III protein secretion systems. *Trends Microbiol.* 12:113–15.

7. Deriving the TTSS from the flagellum by a Darwinian process may be possible (although it is far from certain), since it would involve the production of a less complex structure whose essential mechanism already resides in the more complex structure from which it is derived. Yet that would not help in showing how a Darwinian process could produce the more complex structure.

Appendix D   The Cardsharp

1. Ball, P. 2004. Synthetic biology: starting from scratch. *Nature* 431:624–26.

2. Pawson, T. and Nash, P. 2003. Assembly of cell regulatory systems through protein interaction domains. *Science* 300:445–52.

3. Bhattacharyya, R. P., Remenyi, A., Yeh, B. J., and Lim, W. A. 2006. Domains, motifs, and scaffolds: the role of modular interactions in the evolution and wiring of cell signaling circuits. *Annu. Rev. Biochem.* 75:655–80.

4. Writing of researchers who are trying to manipulate the cell, Phillip Ball (Ball, 2004) observes, "Synthetic biology is the logical corollary of the realization that cells, like mechanical or electronic devices, are exquisitely 'designed'—albeit by evolution rather than on the drawing board."

5. Lim, W. A. 2002. The modular logic of signaling proteins: building allosteric switches from simple binding domains. *Curr. Opin. Struct. Biol.* 12:61–68.

6. Pawson, T., and Nash, P. 2003. Assembly of cell regulatory systems through protein interaction domains. *Science* 300:445–52.

7. Park, S. H., Zarrinpar, A., Lim, W. A. 2003. Rewiring MAP kinase pathways using alternative scaffold assembly mechanisms. *Science* 299:1061–64.

8. Dueber, J. E., Yeh, B. J., Chak, K., and Lim, W. A. 2003. Reprogramming control of an allosteric signaling switch through modular recombination. *Science* 301:1904–8.

9. Ibid.

10. Lenski, R. E., Ofria, C., Pennock, R. T., and Adami, C. 2003. The evolutionary origin of complex features. *Nature* 423:139–44.

11. http://dllab.caltech.edu/avida/v2.0/docs/genesis.html.   This website describes the Avida program, including the setting SIZE_MERIT_METHOD:

"This setting determines the base value of an organism's merit. ["organisms get CPU time (that is, 'food') proportional to their merit."] Merit is typically proportional to genome length otherwise there is a strong selective pressure for shorter genomes (shorter genome => less to copy => reduced copying time => replicative advantage). Unfortunately, organisms will cheat if merit is proportional to the full genome length—they will add on unexecuted and uncopied code to their genomes creating a code bloat. This isn't the most elegant fix, but it works."

Acknowledgments

My citation of results or use of figures from the scientific literature, of course, does not imply that the authors of those works agree with the controversial conclusions of this book. I'm grateful to many people for discussions that clarified the ideas presented in this book. For reading portions of the manuscript in draft form I heartily thank Lydia and Tim McGrew, Peter and Paul Nelson, George Hunter, David DeWitt, Doug Axe, Bill Dembski, Jonathan Wells, Tony Jelsma, Neil Manson, Jay Richards, and Guillermo Gonzalez. I am much obliged to my editor, Bruce Nichols, for his encouragement over the years, and for whipping the klunky draft manuscript into more readable prose. I appreciate the continuing support of the folks at the Discovery Institute, especially Bruce Chapman, Steve Meyer, John West, and Rob Crowther. Far above all, I'm grateful to my wife, Celeste, for her constant love for, and preternatural patience with, a hopelessly distracted husband, and also for bearing, and bearing with, our children—Grace, Benedict, Clare, Leo, Rose, Vincent, Dominic, Helen, and Gerard—who make our house a (very noisy) home.

Index

Page numbers in *italics* refer to figures.

About the Author

Michael J. Behe received his Ph.D. in biochemistry from the University of Pennsylvania in 1978 for his dissertation work on physical properties of sickle cell hemoglobin. From 1979 to 1982 he was a Jane Coffin Childs postdoctoral research fellow at the National Institutes of Health. He is currently Professor of Biological Sciences at Lehigh University. He lives near Bethlehem, Pennsylvania.

9/07